PENGUIN
COMPASS

ASTONISH YOURSELF

Roger-Pol Droit was born in Paris in 1949 and is a philosopher, a researcher at the Centre de la Recherche Scientifique, and a columnist for the French daily *Le Monde*. He is the author of *La Compagnie des Philosophes.*

ASTONISH! YOURSELF!

101

EXPERIMENTS IN THE PHILOSOPHY OF EVERYDAY LIFE

Roger–Pol Droit
Translated by Stephen Romer

penguin compass

PENGUIN COMPASS

Published by the Penguin Group
Penguin Group (USA) Inc., 375 Hudson Street, New York, New York 10014, U.S.A.
Penguin Books Ltd, 80 Strand, London WC2R 0RL, England
Penguin Books Australia Ltd, 250 Camberwell Road,
  Camberwell, Victoria 3124, Australia
Penguin Books Canada Ltd, 10 Alcorn Avenue, Toronto, Ontario, Canada M4V 3B2
Penguin Books India (P) Ltd, 11 Community Centre, Panchsheel Park,
  New Delhi – 110 017, India
Penguin Books (N.Z.) Ltd, Cnr Rosedale and Airborne Roads, Albany,
  Auckland, New Zealand
Penguin Books (South Africa) (Pty) Ltd, 24 Sturdee Avenue, Rosebank,
  Johannesburg 2196, South Africa

Penguin Books Ltd, Registered Offices:
80 Strand, London WC2R 0RL, England

First published in France as *101 Expériences de philosophie quotidienne*
  by Éditions Odile Jacob 2001
Published in Great Britain as *101 Experiments in the Philosophy of Everyday Life*
  by Faber and Faber Ltd 2002
Published in the United States of America as *Astonish Yourself* in Penguin Compass
  2003

10 9 8 7 6 5

LIBRARY OF CONGRESS CATALOGING-IN-PUBLICATION DATA
  Droit, Roger-Pol.
  [101 expériences de philosophie quotidienne]
  Astonish yourself / Roger-Pol Droit
  p. cm.
  First work originally published: 101 expériences de philosophie quotidienne.
  Paris : O. Jacob, 2001. 2nd work originally published : 101 experiments in the
  philosophy of everyday life. London : Faber and Faber, 2002.
  ISBN 0 14 20.0313 1
  1. Philosophy–Problems, exercises, etc. I. Title: 101 experiments in the philosophy of everyday life. II. Title.
  B52.5.D76 2003
  100–dc21   2003040557

Printed in the United States of America
Set in Walbaum Book with Disorder
Designed by Sabrina Bowers

Contents

Introduction:
Everyday adventures

This book is an entertainment. It attempts to indicate essential matters with the lightest of touches. Contrary to what Pascal thought, it is as useless to oppose serious questions, which ought to require all our attention and our energy, as to oppose the futilities that distract us from them. Futility can lead to thought, the laughable can become serious, and depth can succeed superficiality. Not all the time, and not necessarily, it goes without saying. Not every idiocy contains a philosophical pearl.

But there exist ordinary situations, everyday gestures, actions we carry out continuously, which can each become the starting point for that astonishment that gives rise to philosophy. If we are ready to accept that philosophy is not a matter of pure theory, if we accept that it originates in particular attitudes to existence, in the unusual adventures philosophers have had in the realms of feelings, perceptions, images, beliefs, powers, and ideas, then it should not be impossible to imagine experiments to be lived through that may incite further inquiry.

The idea is to provoke tiny moments of awareness. Invent things to do, to say, to dream, that produce astonishment or the unease generated by certain questions. It's about fabricating microscopic starter devices, minimal impulses. Playing on the level of objects.

Each experiment in the pages that follow is to be carried out properly. It is possible to compare them, to modify them, and to invent others. But you must really apply yourself if you are to feel the unsettling of reality that they seek to produce. For that has always been the end in view, since philosophy began: a systematic discrepancy, a step to one side, a change of viewpoint—perhaps a very slight one to begin with—which can reveal the landscape under a quite different angle.

If the entertainment proves useful, it's because it offers such points of departure. Deliberately strange. Even crazy, if need be. But intended always to shake a certainty we had taken for granted: our own identity, say, or the stability of the outside world, or even the meaning of words. What happens afterward will be different for each individual. Different conclusions will be reached. And so much the better. It is enough that the journey has been started.

Obviously, these experiments are founded on certain hypotheses and convictions. They suggest in particular the possibility that "I" is always another, the world an illusion, time a lure, language a fragile veil laid across the unnameable, politeness a moratorium for cruelty, pleasure a form of morality, and tenderness the sole horizon. No one is obliged to share these views. The only thing that counts is that we should be stung, or tickled, into exploring further.

To sum up, the purpose of this entertainment could be contained in this brief exchange:

–Where are you headed? –Wherever you get to!

ASTONISH YOURSELF!

1 Call yourself

Duration:	*about 20 minutes*
Props:	*a silent place*
Effect:	*double*

Sit down in the middle of a quiet room, sparsely furnished for preference. First of all, just spend a few moments attending to the silence, knowing that you are going to speak and to hear. Listening hard to the slightest sounds, think how all this peacefulness is about to end. Prepare yourself for the intrusion of human speech.

Now utter your first name out loud. Articulate distinctly, and then repeat it, insistently. As if you were hailing, from some distance, a person deaf to your calls. Imagine you are calling someone you know, but who cannot see you. From the other side of a field. Or from a riverbank to a boat. Or from one house to another.

To begin with, for the first fifteen or twenty times, you feel as though you are simply calling into space. Calling someone who isn't there—or is inaccessible—in an absurd and ridiculous way. You can lengthen out the vowels or stress the syllables differently as much as you like— nothing doing. But carry on. The door is shut tight.

Little by little you start to get the feeling of being called. Confusedly at first, almost imperceptibly. Hesitant, uncertain. This is where you should concentrate, attentive to the unstable equilibrium between within and without. Go on, keep calling yourself a few dozen more times, mechanically, automatically. It's your voice all right. But it's also that other's, over there. You've only just become aware of it.

Your voice has not become double. Nor have you, obviously. And yet you feel a kind of doubling, as if you have in some sense been internally split down the middle. It's you who are calling, but you don't know to whom you are calling. And it's you who are being called, but you don't know where from. Or rather, yes you do; you know that it's you in both cases, and as for this "you," you presume that it's one and the same. You know it to be so, what's more, and everybody is agreed on this point. And yet it's not what you feel at this moment. You know full well that "you" and "you" are identical, but you no longer feel it fairly and squarely. The one who is calling is the same, and not the same, as the one who is called.

The experiment consists in prolonging, for a few minutes more, this game of within and without, of calling and listening. The point is to feel, from as remote a point as possible, the strangeness of this name that is so familiar, but which you can never use to address yourself without feeling you're someone else. Only other people call you this; you don't, normally, ever call yourself. Go on hailing yourself, at regular intervals, almost shouting out the name from time to time. The aim is to solicit the slight—and not necessarily unpleasant—sense of unease that accompanies those moments when the self becomes a little unstuck from the self. And to maintain yourself in this slightly dizzying situation.

To escape it? To close the gap and stick the edges back together? Simply say, in a clear, loud voice, and as naturally as possible: "Yes! I'm coming!"

2 Empty a word of its meaning

Duration: *about 2 to 3 minutes*

Props: *whatever lies at hand*

Effect: *desymbolizing*

This can take place anywhere, and at any time. Simply make sure, once again, that no one can hear you. Best to avoid the fear of being ridiculed while you're doing it. Speaking to oneself is nothing. But to be spied upon and teased would spoil the desired result.

So, choose a place where no one will hear you. Take what comes to hand, the most ordinary object—a pen, a watch, a glass—or even a piece of your own clothing: a button, a belt, a pocket, a shoelace. Whatever. Just let it be ordinary. Its name is known, its presence familiar. You have always called this object by the same word. Consistent, natural, normal.

Now take this inoffensive, familiar, safe little object in your hand. Repeat its name, in a low voice, as you look at it. Stare at the watch in your hand and repeat: "watch," "watch," "watch," "watch," "watch," "watch," "watch." You can keep going. It shouldn't take long. In a few seconds the familiar word detaches itself, and hardens. You find yourself repeating a series of strange sounds. A series of absurd and meaningless noises that denote nothing, indicate nothing, and remain insensate, formless, or harsh.

You probably experimented like this as a child. Nearly all of us have felt the extreme fragility of the link between words and things. As soon as it is twisted, or pulled, or distended, that link becomes problematic. It becomes contorted, or it breaks. The word dries out and crumbles, a scattered shell of sonorous inanity.

And what happens to the object is no less startling. It's as though its substance becomes thicker, denser, cruder. The object is somehow more present, and differently so, the moment it escapes the fine net of recognizable syllables.

You should repeat this old game of dissociation. Try to observe the moment when meaning dissolves, and how a new, raw reality emerges outside of words. Glimpse the hard scale beneath the prose. Repeat the same word several times, for the same object, and dissipate all meaning. Is it not marvelous? Terrifying? Funny? Just a few seconds are enough to tear that fine film within which we make sense of things, smug with the power of giving things a name.

3 *Look in vain for "I"*

Duration: *indefinite*

Props: *none*

Effect: *dissolving*

It's one of the terms you employ most frequently. During the day, the word "I" crops up in nearly all your sentences. Since your tenderest childhood you have ceased referring to yourself by your own first name. "I" has become the word by which you express your desires, disappointments, projects, hopes, acts of all kinds, physical sensations, illnesses, pleasures, plans, resentment, tenderness, your weakness for vanilla, and your aversion to fennel. For a long, long time you have linked this tiny word to your multifarious mental states. It is intimately involved in your feelings and your memories. Apparently, nothing is possible without it. It is there in all your stories and all your judgments. Not a single decision, not the slightest rumination escapes it.

The curious thing is: everyone uses the same word. The most irreducible intimacy, the most singular existence, for each one of us, is linked to a word that we neither chose nor coined, and that everyone else employs in exactly the same way. A pronoun in the language. There's nothing less personal than this "personal" pronoun. The particular existence it refers to remains, linguistically speaking, completely interchangeable. It could be anyone who says "I'm happy" or "I'm sad." All of us, in all our difference, refer to ourselves by exactly the same word as everyone else. A most paradoxical situation. But you don't think about it, and nor does anyone else, of course. You have enough to do without worrying your head about questions of that order.

And yet, try to pin down this "I." Does it exist? How can you find it? What does it look like? If you apply yourself to asking these questions, and trying to resolve them, you'll find that this "I" is neither simple to localize nor to authenticate.

This is not a brief experiment, whose limits are easy to circumscribe. It can come to seem, on the contrary, like a long pursuit. You need time, different occasions, a certain application, and stubbornness. So where is this blindingly obvious "I?" You will seek for a long time, in different places and under different aspects. And there is a strong chance that, at the end of it all, you'll return somewhat at a loss. Which is where things start to get interesting.

Among the avenues of inquiry you might like to pursue, it's worth remembering the existence of the body. Is not this "I," which is both individual and yet assimilable to others, in fact identical with the body that houses it, with its habits, its weaknesses, its vulnerabilities, and its particularities? But there's no trace of an "I" in your body. Not one of your cells lives longer than ten years. No part of your body has persisted unchanged. So what will you address as "I?" The form? The ensemble? The general organization? There remains, famously, the phenomenon of thought. All may change, but not your memories, not your sense of remaining unchanged despite corporeal alterations. But even here, you cannot put your finger on an "I." All you will ever discover are thoughts, sequences of thoughts, memories, associations of ideas, desires—all of them pressed into service by what you call your "I."

To all these sensations, to all these mental events, the "I" seems to provide a common denominator. But it neither supports nor drives them. It merely imparts to them something like a family resemblance, a shared aspect to what are very diverse notions and feelings—something like a color or an odor. A way of seeming, a style. Nothing more. "I" is not a someone or a something. And yet neither is it just a word. It must be a refrain of the self, a secondary quality, at one remove.

If you manage to carry the experiment thus far, you will need to know what to do about this sensation. What impact will this impossible discovery about your "I" have upon your existence? How will you cope once your "I" has gone missing? That is another story.

4

Make the world last twenty minutes

Duration: *21 minutes*

Props: *a world and a watch*

Effect: *terrifying or reassuring*

The past clings on. It is present in the smallest actions. It coils itself around our thoughts, even in those that seem unconcerned with it. The future also ceaselessly sustains the smallest of our projects. It accompanies our slightest expectation.

What would happen if we tried to rid ourselves—in a spirit of illusion and of play—of these terrible constraints? Imagine, therefore, as far as it is possible to do so, that the past never happened and that the future does not exist. Let us believe that the world, this world, lasts only twenty minutes. It was created from nothing, just an instant ago, as it is, and us with it. One minute earlier, it did not exist. Everything the world currently contains by way of relics, ancient ruins, libraries, monuments, archives, distant or recent memories—the whole lot—has just materialized, at the same instant. The archives are there all right, as are the witnesses to a past, but the past they speak of never existed—until a moment ago.

This world—infinite, diverse, multiple—has a life

expectancy of exactly twenty minutes. Beyond which time it will disappear completely and definitively. Not in some gigantic conflagration or cosmic explosion. Not in some terrific fire or furnace. Just a brusque extinction. Like a soap-bubble bursting, or a light going out.

Make yourself at home in this twenty-minute world. Remark the extent to which it is, in a sense, identical to our own: same dimensions, same skies. No object is any different. The same people are doing the same things. And yet: it is not at all the same universe. A world that lacks the depth of a real past and the perspective of a viable future may certainly seem completely identical, but it still differs radically from our own, due to this time limit. Before this ephemeral universe has completely disappeared, try hard to understand—you who were under the illusion that another reality existed and will exist—to what extent your thought process is habitually different from this existence which is even now counting down. The more you experience this contrast and this distance, the more you will feel the importance, for us, of an immemorial past and a distant future.

As the fatal twenty minutes approaches its term, you should feel, furtively, the dumb terror that everything will, effectively, disappear.

Most likely this will not happen. You can then emerge, at the twenty-first minute, from this objectless terror. Now concentrate on savoring your relief that the world goes on.

Later you might feel, like an aftertaste, a secret disappointment that nothing was obliterated.

Bad loser . . .

5

See the stars below you

Duration: *30 to 60 minutes*

Props: *a starry sky*

Effect: *cosmic*

Preferably a summer night. There should be no clouds. Even better if you have a garden, where you can lie out somewhere dry, and with an hour or so free in front of you.

So there you are on your back. You look up at the stars, their infinite number and their far-flung strangeness. You must try to feel both awed and relaxed, taken up into the mysterious and reassuring presence of the night.

The scene bristles with clichés: the milky softness, the warm darkness, the twinkling that makes us feel so small. All the conventions are in place. Above all, don't hold back; plunge right up to your neck in them.

Take the necessary time, and wait until you get the feeling you are riveted to the ground, almost crushed by that immensity, a tiny dot with all of infinity above you.

The experiment consists simply in overturning the universe. Little by little you will now convince yourself that the stars you are watching are below you. You are overlooking them. A massive force keeps you on the earth. But the vast sky is down below. You are flying over that abyss of stars, into which you risk falling forever.

It won't work immediately. Some time for adaptation is necessary, and a floating state of reverie is better than a conscious effort. The process is similar in kind to looking at three-dimensional drawings. For a long time you remain there looking at a flat sheet covered with signs that look not only flat but incomprehensible. You have to put up with the waiting. And then suddenly, the whole thing opens up.

You really feel that everything is below you.

It would take nothing, a sudden gust, a brief failure of gravity, possibly even a momentary lapse of attention, and there you are, floating off very slowly, between the earth and nothingness, traveling down the sky.

When you get up, do so slowly—and mind your step.

6

See a landscape as a stretched canvas

Duration: *20 to 30 minutes*

Props: *a gentle landscape*

Effect: *surprise*

The sea or the countryside, rather than the city. A landscape that is relatively simple, not too tortuous, and almost uniform. Few contrasting colors and few conflicting forms.

You settle down to look at it. Don't stare. Don't scrutinize. There's nothing for your eye to seek out, and it should avoid stopping at any particular point. On the contrary, let it glide over the whole, disengaged and slightly vague. As though nothing could halt or attract it, no angularity, no particular form. The end result of all this is that everything must seem to you a single surface, flat and without relief–like a painting. The time this takes can vary. Sometimes it happens very fast. Everything depends on you, on your mood, and on the landscape.

When you come to perceive the view as made up of a single smooth surface, devoid of internal tensions, the experiment can really begin. Imagine that everything you see, from earth to sky, whether still or in motion, is just a detail on an immense, stretched canvas. On a giant screen, a "total" screen, shown in perfect focus and definition.

If you get to this point, and you have become sufficiently convinced that all you have opposite you is a colored, scarcely moving canvas, or a gigantic, old-style cinema screen, you can now begin to imagine this canvas being folded up. You are about to see this great curtain, which contains the entire landscape, reveal something behind itself, as, very slowly, it starts to fold.

In which direction? Up? Down? Diagonally? Or horizontally, from one side? That will depend on your mood.

The point to reach is that moment when you can start to believe in the existence of a world that can fold away. You must try to prolong the mild apprehension you feel as you wait to discover what lies behind. Don't necessarily picture some abysmal darkness, some gulf or fiery furnace. Nothing like that. The important thing is to feel how the world is always liable to slippage, to disappearance, to an absence of certainty. And that you should feel this mild apprehension in front of any landscape.

Before the curtain goes up, you can quit the experiment by announcing a five-minute interval. Or you can try, but you won't get out of it that easily. From now on, you know that the curtain won't wait on ceremony for you. The solidity of the real has been diminished, its self-evidence touched with corruption. Anything can happen, everywhere, all the time, spontaneously.

7 Lose something and not know what

Duration: *unpredictable*

Props: *anything*

Effect: *anguishing*

Most things one can prepare for, but never for loss or memory failure. For this reason, the present experiment cannot be set up. The two essential conditions must come together by accident.

You need to have lost an object, trivial or important, and you need to know that you have actually lost it, but be unable to quite recall what it is you have lost. You need to have fallen victim—and this is rare but not impossible—to a double loss: of an object and of a memory.

You know you have misplaced one of your ordinary possessions, or something that someone has entrusted you with, or even some document that was in your care, no matter . . . and yet you still don't know what.

For the moment, you merely have the confused feeling of a gap in the continuity of things, without knowing how to identify it by focusing it more sharply. An irrecuperable malfunction, an anxiety redoubled for not knowing what it is that you have lost—this is the situation you need to encounter.

Let me repeat that this happens rarely, and is impossible to provoke. One can only wait for it to happen. And prepare for its happening. After all, perhaps it happens more frequently than we think. Usually, we tend to hide these moments from ourselves. We bury them, and smooth the sand over them. They mingle with the glittering dust of our routine daily acts, like the suspended motes you can only see in rays of sunlight.

Here, however, you must await such moments and then pay attention to them. If such a rare occurrence does happen to you, then the experiment will consist in settling in to the strangeness you will feel. It is not regret because there is no content to warrant that emotion. Nor is it shame, or a kind of generalized embarrassment. It is both vaguer and more terrible. To have forgotten a forgetting and yet to know, "confusedly," that a forgetting has occurred. But what is this confused knowing? Is there such a thing? What do you call it? How can this strangely slanted vision of time exist? Like seeing oneself from the outside, but from one side only, distortedly and at an angle, and with weak vision.

At this point you may be seized with that blank fear of some irreparable loss, which suddenly opens up without you knowing in exactly what it consists. To find out more, you may consult books on psychosis or mysticism— the choice is yours.

8

Recall where you were this morning

Duration: *variable*
Props: *none*
Effect: *suspenseful*

This is an experiment for the overworked. Suitable for exhausted travelers, pressurized salesmen, stressed-out decision-makers, for everyone who overdoes it.

It works best if it is carried out late in the day. Or if the day has been especially chaotic. Or if it follows several days of overwork, movement, and excitement. In short, do this experiment after a series of exertions have reduced you to a state of almost drunken fatigue. Choose the moment when you think you are close to losing control, and are starting to forget details. The effort needed to hold together so many choppings and changings and so much data is starting to come undone.

Once you've reached this state of extreme nervous tension in which you wonder if you can go on, the experiment becomes very simple. Ask yourself: where was I this morning? Variants are possible: what was the first sentence I heard? What was my first meeting? With whom did I spend the night? (And so on, according to your lifestyle.)

Multitudes of people will reply to these questions without a second's hesitation. They know instantly where they woke up. What they ate, said, read, heard, the people they've seen. Such questions are without interest for those who live according to a strict routine, with its monotonous hours and unchanging days. Such people know at once because today was the same as usual, the same as always. The office, the shop, the farm, the factory. No change there.

For the others—the stressed-out nomads, the overheated mutants—the thread is less easy to rewind. When meetings, decisions, and rendezvous succeed each other, knowing exactly what one was doing a few hours before can prove very difficult. What's important here is not the failure of memory, or the feat of memory. It's the sense of hesitation. Prolong that moment. For a few seconds, a few minutes. You remain suspended, hesitant, no longer knowing what, in your own existence, has only just preceded the present moment. You know very well that your body was somewhere, you're certain that it remembers where, and that the right answer will come. And yet the moment lasts, continuity has been broken, you remain somewhat apart from yourself, at a distance from your own time. You know perfectly well it was you: that that moment occurred, that you heard that sentence, that you woke up. And yet, at least for now, the memory does not come back; you remain on the edge of the instant—there are gaps in the past and they worry you. Because continuity is also a matter of faith, evidently.

9 *Hurt yourself briefly*

Duration: *a few seconds*

Props: *none*

Effect: *back-to-earth*

You are bored. The play is interminable. The lesson is without interest. Or you're waiting for a phone call that doesn't come. Or you don't know what to do next, and you are in two minds. The world is veiled in a kind of mist. You feel you are becoming inconsistent yourself, as if your substance had begun to lose definition and to spread out vaguely all around you. As if you are becoming increasingly vaporous, milky, and weightless. You no longer know exactly who you are, or where you are. Boredom has started to dissolve you.

Pinch yourself. Hard. Where it really hurts. The inside of your arm, your neck, or your groin. The pain caused must be brief, but intense. Enough to make you utter a cry, which you may well have to smother. To outwit your defense mechanisms, act quickly. Allow yourself no time to anticipate or prepare for the pain. Be sudden. Try to take yourself by surprise, so to speak. Do everything in your power to hide your intentions. The pain must traverse you as though by accident, like a sudden collision. It must descend on you, like a lightning flash in the middle of the torpor of the day.

If you are sufficiently violent, the effect is certain: you recover reality, your body is returned to you, you know where you are, the mist dissipates, you emerge from your boredom, you return to the world.

Just one question remains, which you should ponder: why should the experience of pain return us to reality? Is it a simple reminder? The effect of contrast? Or have we, in the course of our millennia, created such a way of life for ourselves that pain has become the first symptom of the world? A piercing question.

10 *Feel eternal*

Duration: *limitless*

Props: *none*

Effect: *restful*

Imagine that our eternity is not a matter of faith. It is a fact. Or at least envision it as a perceptible reality, not demonstrable by reason. An endless series of abstract arguments will never convince us. To feel eternal, we have to try it out. Which may sound crazy. But try it just the same.

Imagine the journey toward the perception of the eternal as a journey toward the inside of your body. Your skin exists within time, it is on the periphery, the outer circle. The heart is also in time, beating away, and so are the stomach and lungs, working according to their respective rhythms. It is below, and further in, that space out of time is to be found. In this pure space, hidden from your gaze, you will contemplate the unpeeling of the skin of time. You will watch it detach itself from you and from everything around you, like an empty shell rolling away into the distance.

If you were in that space, you would see your own thoughts succeed each other without pause, and leave no trace; you would see all things move in the present, in a present dilated, enlarged and extended to the dimensions of the universe.

The experiment consists in living the superficial nature of time from within. You have to feel, first as a kind of unusual dizziness, then as a truth that is more and more familiar, that the most fundamental kernel that constitutes you has nothing to do with the successive events of time. You contemplate them. You accompany them. But you are not a part of them. That at least is what you must persuade yourself.

It is not a question of deciding whether or not things are really thus. What is essential is that you should have, however fugitively, the sincere impression of its truth. No matter if, in actual fact, we are ephemeral beings. If, at the heart of this incessant flux, this endless and discontinuous march of the hours, we once experience our eternity with total conviction, then we shall have escaped time. The illusion is enough.

There is nothing more to say about this experiment. The real difficulty consists in trying to understand it. And in persisting until we achieve clarity.

11 Telephone at random

Duration: *20 to 30 minutes*

Props: *a telephone line*

Effect: *humanizing*

Lift the handset. Start dialing a number, any number, blindly. Press the numbers at random. Wait and see what happens. To begin with, the experiment is mostly disappointing. Engaged signals, recorded error messages, silences, blanks. Dead ends. Unless you're very lucky, your first attempts lead nowhere. The telephone does not work at random. So you must arrange matters to increase your chances of success.

Begin by determining the total number of figures you will dial, which varies depending on the country you're in, the requisite codes, the region you are thinking of contacting. You can either limit yourself to national calls or extend your luck to the four corners of the world (depending on your mood, your languages, and your budget).

Clearly none of this should be treated as a prank. The game we are playing has nothing in common with the kind of practical joke played by adolescents the world over. And this, indeed, is the first thing you need to impress upon your interlocutors. "I'm phoning you at random. Can you tell me who you are?" You must get them to agree, if you can, that it's no joke.

What happens next is unpredictable. They'll slam the phone down on you, or you'll begin an unlikely conversation with the receptionist at a steel girder factory in Manchester. They'll insult you, or else a strange, semi-anonymous relationship will be initiated with someone who was a perfect stranger a moment before.

The experiment is not about making new friends or chatting people up from the comfort of your own home. Not that there's anything wrong with that, but it's not the point. Rather it's a way of experiencing the *density* of the human world, both close at hand and far away. Telephoning at random ought to be the starting point of micro-adventures into this density. Of infinitesimal odysseys. Instantaneous disorientations, sudden abyssal faults in the daily routine, little pockets of strangeness. To return to earth, just hang up. But it takes a moment to adjust. Strands of otherness still hang in the air. Or you've left some thread of your own behind, and you don't quite know where.

12 Rediscover your room after a journey

Duration: *10 to 20 minutes*

Props: *a return home*

Effect: *restful*

You need to have come back from far away. Or else to have been away for a long time. You have got out of the habit of your usual routine. You have slept in different beds, and become accustomed to different foods. You have lived through a change of climate, rhythm, and horizon. You have heard other languages and engaged in different activities. Your body and soul have shaped themselves to new habits. And right now your own front door is approaching. For several minutes, landmarks have been reinstating themselves. This is the moment to explore. You look around at the streets and houses of the neighborhood. You know very well they are as they were. And yet it's not quite the same as before–though it is difficult to say exactly what has changed. Nothing, clearly. Something, nevertheless. And not simply in you. It's in the objects themselves, or between them and you, that something seems to have come unstuck.

You open the door. Go straight to your bedroom. Lie down on the bed, and look around you attentively. You must first repossess the volume, reframe the distances, and readjust the colors. None of these words is right. The process, which unfolds very fast, is much subtler than the available vocabulary allows for. You know this space by heart—its shape and its colors. But you have not seen it recently. You have had to get used to other sensations. And now, recovering these old ones, you are aware of their familiarity and also of the distance you have traveled from them.

Note carefully anything you might actually have forgotten. Tiny details: a stain on the wall, a fold in the carpet, a slight irregularity in the floor, things like that. You know them. But you hadn't thought about them, and they had faded from your mind. Without astonishing you, they surprise you just a little. Try to remain in this state of unstable equilibrium, between the old grooves, into which, soon enough, you will slide back effortlessly, and the various adjustments you made during your absence. This time-in-between is of very brief duration. Soon you will paste all the pieces back together, and recount your journey in the past tense.

Before your days and duties resume their unquestioning course, ask yourself how can your room have waited for you, and how can it not have changed? It's hard to understand how a reality can remain unchanged in your absence. And you, how have you contributed to the fact that your room has remained identical? Have you carried it with you in a corner of your memory? Have you supported, nourished, animated it? Is it to you, to itself, or to someone or something else that this room owes the fact that, when you weren't there, it refrained from collapsing into nothingness, or reemerged from nothingness intact?

Naturally, some people will shrug their shoulders at stupid questions like these. Things stay put and that's that. And we come back to them. There's nothing more to it. Are you so sure?

13   *Drink while urinating*

Duration: *1 to 2 minutes*

Props: *toilet and glass of water*

Effect: *wide open*

For hundreds of thousands of years the vast majority of humans have lived and died without trying the following experiment. It is, however, both extremely straight-forward and extremely interesting.

Like everyone else, you urinate. And at other moments you drink. What you do not know is what it feels like to do both at the same time. This experiment will show you.

So, just have a large glass of water at hand. When you begin to urinate, start drinking. As far as possible, you should try to drink the water straight down, without pausing. You will feel quite bizarre sensations almost immediately. The water you evacuate seems to be synchronized with that entering your mouth. You will then visualize, and above all feel, your body to be organized in a way which until then you had never imagined possible. The water you are drinking seems to exit directly from your bladder. In a few seconds you will feel directly wired, from throat to urethra, from stomach to bladder–

a physiology that is impossible but that you intuit, directly and unquestionably, to be real.

It has taken no more than a few moments for you to discover this wonderfully simple body, and you feel there can be no other. No more intestine, no kidneys, no filtration time, no waiting. Water pours through you vertically, a cool liquid washes through you in a peculiar and palpable way. Your system seems to have opened inside out, with the water flowing smoothly from inside to outside. It is like—take your pick—the cosmic flux or an automatic washing machine.

This experiment, which can be repeated indefinitely, which costs nothing, and which is likely to procure ever new sensations and surprises, has not hitherto been considered a thermal cure.

14 Make a wall between your hands

Duration: *about 10 minutes*

Props: *none*

Effect: *doubleness*

Put your hands palm to palm, fingers against fingers, level with your eyes. Separate your palms, with the fingertips of both hands still joined. Now and then move your palms toward each other, without letting them touch. You push, and resist with each hand, pressing the fleshy part of your fingertips harder and harder together.

Each of your hands must take turns opening and closing, as if you were trying to push back a wall, or the pressure of some flat, inert surface. Flex your finger joints as fully as possible, feel the tension in the muscles of your palms, the stretching of your ligaments. If you repeat this process of pushing and flexing several dozen times, you will discover the ambiguity of the situation.

You are both he who presses and he who resists. With your hands on either side of you, you will experience an unusual difficulty in knowing where you are. You are the other, the other is you. And it gets weirder if you prolong the experiment. You can no longer tell which is the living element, and which the inert. For each hand, the resistance to its movements constitutes the outside, and feels like a wall. And on each side you feel the living effort, not the flat surface. You feel the skin, the flesh—but not the presence of the wall. That hypothetical, virtual, and yet palpable wall—which you cannot place exactly.

The situation is complicated by the fact that your hands are level with your eyes: what you see does not coincide with what you feel. The visual image is quite normal—two symmetrical hands. The feeling is quite abnormally asymmetrical: each hand is the living contradiction of the other. In this hand-to-hand combat with yourself, a kind of closed circuit *mano a mano,* you experience for yourself that "I is another."

15 *Walk in the dark*

Duration: *a few seconds*

Props: *a dark room*

Effect: *disorientating*

Suddenly it's pitch dark. Power cut, or sudden awakening, or attempt to avoid waking those who are asleep . . .
The reason doesn't matter. You are walking in the dark.
Preferably without expecting to. No light to give you your usual bearings as to obstacles and distances. With only your memory to guide you, you must cross a familiar room, your bedroom or your sitting room, in total darkness. What is worth exploring here are your moments of uncertainty. Your gropings seem to suggest that you don't know how to navigate the familiar space you've crossed a thousand times. How many steps are there between the bed and the door? Is there nothing between them? Where's the arm of that chair? The corner of the bed? These reassuring places bristle suddenly with question marks.

The simplest movements become fraught with risk and perplexity. Worst of all, you can no longer judge distances. What you thought you knew, in the light, has become uncertain. Nothing is guaranteed. You stretch out your arms, thinking you're about to bump into something, touch the wall, or brush past the doorframe . . . Nothing there. You keep groping in the void. Almost from the start, what invades you without your wholly realizing it is in fact the benightedess of ignorance. The darkness makes you stupid. It has thickened your head and destroyed your bearings. Suddenly you bump into the corner of the chest of drawers. You hadn't imagined it was there. So you were completely wrong in your calculations. You were not where you thought you were. The chest has loomed out of the darkness and struck you a calculated blow, high up on your thigh, just where it hurts the most.

The absence of light skews all your estimations. It confuses your contours, and your body seems uncertain and at a loss. You can only move in limited fits and tiny starts. And yet, very little is actually missing from your picture of things. Known reality is still unmoved and in place. Nothing has budged, neither the objects nor the relations between them. Nevertheless, they have become incomprehensible. Distanced and vaguely menacing.

In the dark, the world is supposed to be "the same" as in the light. But you have only to test this proposition to find that the world changes completely, depending on whether it is visible or not. What we call "the world," "reality," "normal life" reposes inside a thin, easily disturbed stratum.

16 *Dream of all the places in the world*

Duration: *20 to 30 minutes*

Props: *none*

Effect: *joyous*

You're tired of being where you are. The place is limited, repetitive; it holds no more interest for you, let alone surprise. You are seriously fed up with this one place, forever identical with and closed in upon itself. And yet escape is close at hand. Just think of the infinite variety of places that exist, near or far, at this moment.

Famous sites: the Piazza San Marco in Venice, the ramparts of Jerusalem, the entrance to Central Park off Fifth Avenue, the Yamoussoukro cathedral, the Pyramids, the Little Mermaid in Copenhagen, the Plaza de Mayo in Buenos Aires, the Colosseum in Rome, the Champs-Elysées, the Forbidden City, Beverly Hills, Red Square, the Parthenon, Trafalgar Square, the Red Fort in Delhi, the Topkapi . . . an endless, infinitely extendable list of styles, squares, buildings, cafés, statues, panoramic views of every kind.

There is more. Think also, until dizziness sets in, of the infinite multiplicity of humbler places. Anonymous, unpresuming spaces: backyards, little squares, impasses, narrow streets and alleyways. And even of these humble corners: storerooms, barns, larders, cellars, cupboards, garages. In the humidity of the tropics, the dryness of the deserts, the damp colds of the misty regions. With palm trees or birches, cacti or ancient pines, white sand, red rocks, mud, permafrost, the pure whiteness of the wave against the deep ocean blue.

And then there is the inexhaustible list of what people are doing, what everybody is doing, at this instant, in all the places of the world: making love, crying with pleasure or with pain, eating, dying, sleeping, sweating, toiling, enjoying themselves, amazing themselves, envying each other, traveling, cooking, reading, returning home, singing.

You can bathe in this multiplicity, be carried away by this infinity of alternatives. The place you are in isn't simply one spot within an infinity of others. It contains all the others. And it's all entirely within your head. Permanently available. To one and all.

17 *Peel an apple in your head*

Duration: *20 to 30 minutes*

Props: *none*

Effect: *concentrating*

In general we consider ourselves capable of reproducing everyday reality with a certain degree of exactness. Familiar objects or places, foods, repetitive actions are easily summoned within our minds. We imagine that we can switch on the screen of our mental cinema (so to speak) and project upon it—with a degree of precision—all these well-known images. We perhaps experience greater difficulty in summoning up noises or, especially, smells. And reliving the sense of touch (a light caress, a kiss) is probably harder still.

And it may be that, in spite of everything, our confidence in being able to reproduce external reality in our heads fairly easily and efficiently is to a large extent misplaced and illusory.

To test how difficult it is, try peeling an apple in your head. The exercise seems simple. You visualize the fruit, the knife, the incision, the peel, and there you have it. And yet . . . To ensure that your image bears some relation to reality, you must first choose a variety of apple, visualize its exact size, color, texture. Your apple must be of a particular variety, but it must also be a particular apple, with its own range of colors, some parts darker or lighter than others, and with its own marks, stains, tiny wrinkles–all these have to appear to you with absolute clarity. Imagine the knife: is its handle made of wood? Plastic? Metal? Shiny? Is the blade smooth? Dull? Sharp? Is it a kitchen knife, the family silver, or a fancy Henckelf?

And now, how will you proceed? By trying to peel it in one go, producing a perfect spiral of peel, by turning the apple in one unbroken rhythm? Or by cutting it into quarters and then peeling the skin off all four? In each case you must visualize your movements with clinical precision and photographic accuracy. The goal of the experiment is reached when this documentary film of peeling unrolls in your mind shot by shot, image by image, second by second. Without a single stop, a single rubbing-out or mistake. No blurring and no hesitation. Above all, no blanks and no repetition. And you are not allowed to splice together two sequences.

You won't manage it, or not without considerable practice and mastery. Most probably you'll lose the thread. The apple changes color or form, its characteristics become unstable, the peel doesn't fall as it should, the knife goes askew, the movement becomes jerky, the images jumpy, difficult to join seamlessly to each other. If you repeat the experiment several times, you'll note an improvement in the result. Progress is possible, albeit slow and sometimes painful. If nothing else, it's a good exercise in concentration. But this experiment underlines first and foremost how essentially unfaithful the mind is to reality, how difficult it finds retaining or reproducing that reality correctly, and how presumptuous it is when it imagines it can.

18 Visualize a pile of human organs

Duration: *30 to 40 minutes*

Props: *anatomical plates (optional)*

Effect: *pitiless*

The starting point is simple. A living hand on a human body has nothing anguishing about it.You don't even notice it. Or if something about it does draw your attention (its delicacy and subtlety or, on the contrary, its squat, square, stubbiness) it does so in a meaningful, living way, instantly accompanied by a whole series of associated ideas. By contrast, a hand that is inert, detached from the body, and isolated will first of all give you pause. Even when the hand is famous, and cast in plaster (the hand of Voltaire or Chopin) the effect can be shocking. Which was why the prisoners on Sakhalin sometimes chucked a hand–amputated from one of their fellows–into cargoes sailing out, to tell the world of their existence.

The effect is still other, and much worse, when hands are piled up (you need only imagine hands in plaster, cardboard, or wood) on top of each other like a heap of inanimate things. When fragments of human bodies, all of the same sort, are stashed high pell-mell, like objects deprived of their function, nameless stumps, one experiences a very particular unease. Nothing like a butcher's displays. You may walk through a big meat market with some disgust and a sudden lassitude, but this monotonous exhibition of dead meats at least has some meaning. There is nothing unusual here.

But when you imagine a heap of hands, or arms, or a pile of feet, you will not know how to react. Because each hand or foot calls for a body and demands to be reattached and replaced in its normal context. And also because the heap itself seems to freeze them in their isolation. To the absurd detachment of the organ is added the horror of its being among a thousand similar rejects. It is reasonable enough that screws, say, should be stocked with others of the same size, and the same is true of any category of object. But nothing can humanly explain a heap of human organs.

Failing the chance to contemplate such images in reality, you can always try to hallucinate them. Imagine, just in front of you, a pile of a hundred legs, none of them a pair, of different colored skin, of diverse ages, plump and round and wrinkled, small and big with varicose veins, hairy and hairless, red and pink and bluish, facing in every direction. Look at the toes, at those with gaps between them or at those all squashed together, the broken nails, the swollen ankles, the knobby knees.

You can repeat the experiment using just fingers, or shoulders, or breasts. You can try and take it further, using hearts, lungs, and livers, but the result—while perhaps more horrible—will be less disturbing than if you use heads, disfigured faces, eyes closed or open, blue lips, and matted hair. You can then conceive of a world where the human species, no longer living normally, will have been dismembered and stashed in piles at every crossroads—organs in heaps along every roadside to signal the triumph of a new order.

19 *Imagine yourself high up*

Duration: *15 to 30 minutes*

Props: *a closed room*

Effect: *ascendant*

You are anywhere down below. At sea level, or just above. The experiment consists in trying to levitate everything around you to a considerable height, say to 4000 meters. This is pure auto-suggestion. Nothing in what you see around you will be drastically modified. A closed room is best, preferably with the curtains closed as there must be no view of the outside.

You simply have to rise little by little into a light that is thinner and more transparent. Your breathing will become deeper and faster: there is less oxygen. If possible you feel a slight itchiness in your nostrils. Your temples throb intermittently. You may experience slight dizziness and your head may start to swim. Above all you will try to induce a slight feeling of pressure around the heart. It is almost permanent, and gets worse if you move. You will notice that your movements are visibly slower, and your actions more sluggish. Your thoughts are less well articulated.

You may not manage all these effects on your first attempt. Try again. With repetition, results can show a sharp improvement. With sufficient practice, you may expect to produce this altitude-effect almost faultlessly.

The question remains, why? The impressions thereby induced are relatively unpleasant. The apparent benefit is nil. At first sight the change in altitude offers nothing new visually. It gives no particular access to realities blurred by surface appearances. So what's the point? Why induce hallucinations, why strain to produce so many false effects? Because, for one thing, you'll acquire a few doubts as to the existence of objectivity. And for another, you'll come to this conclusion: that it is possible, at least at certain moments, to dream the world using one's body. And that, after all, is no small matter.

20 Imagine your imminent death

Duration: *5 to 10 minutes*

Props: *none*

Effect: *lightness*

At any moment we can die suddenly. Think, therefore, as so many of us do, how you risk death when you take a plane. Or when you set out on a long car journey. Or when the train pulls out. You can also get killed by a bus, a truck, a car, even a motorcycle. Implausible accidents lie in wait for you everywhere and at every second. In fact, when you think about it coldly, you have no reason *not* to fear imminent death. If you choose to discount such a hypothesis it isn't just because the thought is none too pleasant, but above all because the probability of it actually happening seems very slim. And rightly so. Your chances of still being alive in an hour (and even tomorrow) are relatively good. So why worry about an improbability?

The only snag is that your death is a certitude. Not the day, not the hour, agreed. But it is absolutely ineluctable, guaranteed, without fail, and without exception. You must therefore imagine your own death, its necessity. Try to visualize your deathbed agony, your corpse, your burial, your rotting body, your skeleton. Visualize the tomb with its horrible liquids. Understand that you will never see the light or the curving earth again. You'll have finished forever with its warm winds, its wetness, its flashes of color, its scents. You'll know nothing of flesh, to caress or bite into.

It may be that you find these ideas upsetting. You will doubtless be relieved to know that your distress is absurd and, in fact, without foundation. You are dead, otherwise you wouldn't be buried and in the process of rotting. At the same time you're still alive, and capable of being affected by feelings and emotions. Therein resides your error. These images exist in your head now, and in your living body. When you're dead, they'll no longer exist.

We cannot imagine ourselves dead. Thinking about it can only ever be an activity for the living. Your whole imagination belongs with life. However morbid, sepulchral, vampire-ridden, however full of cobwebs and coffins, it has, strictly speaking, no link with death. It has nothing to do with it. There is only one universe. It has no outside. The thinking we do about our own outside is done from inside us, and tells us nothing about that outside. Does that reassure you? Clearly not, but you will have glimpsed one of the differences between life and philosophy. The former panics, is moved, gets impatient and agitated. The latter remains convinced that everything will work out if only we think about it in the appropriate way. Which is false. Or at best approximate.

21 *Try to measure existence*

Duration: *a lifetime*

Props: *rulers, weights, tensiometers, particle accelerators, etc.*

Effect: *frustration*

The world used to be of different lengths. Notions of weight varied according to region. So much so that it was impossible to say exactly how much a loaf of bread weighed, or give the exact dimensions of a door. Daily life was fairly rough and ready in terms of exact measurement. In principle, the world had been uniformalized mathematically, but in reality a wide margin of error and uncertainty subsisted.

We have changed all that. Norms have been fixed and standards uniformalized. We never stop measuring the things around us. To make a cake, ingredients must be measured. Decorating a room, repairing an engine, constructing a model to scale, planning a vegetable garden— all these activities require the taking of measurements and some basic math. Rightly, you trust to them more than to your own guesswork. There are no journeys without maps, signs, flight paths, sextants, compasses, altimeters, tachymeters, satellite links, radar, global satellite positioning, and whatever newfangled machine comes next. You measure your children, you weigh and analyze them. And you yourself are regularly submitted to similar computations: blood tests, urine tests, stool tests, sperm counts, cell counts, skin samples, radiology, biopsy, endoscopy. Measured, weighed, and tested from every angle. How much carbon dioxide do you exhale, how much potassium and albumin do you evacuate when you urinate, the amount of fat and sugar pumping through your veins. Others worry, you yourself worry, about your weight, your arterial tension, the glucose in your blood.

These reckonings are all useful. But you should also try the mental experiment that shows these exercises to be essentially secondary and vain. You might inquire, for example, how to measure existence. With what instrument? And in what units? Following which code and by what bearings? Would you say that your existence is adequately measured in yards covered on foot, in miles traveled by car, in years, in days, in hours, in seconds; by heartbeats, by liters of sweat, urine, blood, in pounds of flesh, potatoes, or meat, in liters of wine, in pages written, in time wasted, in love given or received? How can it be measured?

Numbers cover the world and tie down reality. Life can in fact be described by a series of equations, a dense grid of size, mass, and energy. But none of this can measure existence.

22

Count to a thousand

Duration: *15 to 20 minutes*

Props: *none*

Effect: *critical*

Initially, nothing much new here. Counting to a thousand will take a certain time (around fifteen minutes, or nine hundred seconds) and ought to be monotonous. It seems predictable and regular. What you expect is a flat, mechanical exercise.

Not so. Serious turbulence is inescapable. There are easy parts, downhills, long straights like the old main roads lined with poplars or plane trees, and then there are hills, escarpments, sudden right-turns, especially when you come to the transversal ranges of the five-hundreds. You expect nothing but figures, and here you are back in your childhood, at primary school, among inkwells, work-smocks, sponges in satchels, the playground. You're back in the Russian mountains, among the "scenic railways," and the black marks for bad behavior. You're counting in black and white.

What ought to be a routine mechanical operation has become an altogether more difficult and complicated affair. Have I missed out a ten? A one, or a hundred?

Didn't I just make a mistake, when I was thinking of something else? Rather than being easy, ordered, and continuous, the distance from one to a thousand is rutted and potholed. There's always a risk of you getting bogged down for good, or falling into a gap. Of hesitating, losing track, and starting all over again. Will this go on forever?

No, you've reached the end. What have you learned? Just this: that one thousand is a big number. You can run through it, but it takes time—a good quarter of an hour—complete with ups and downs. You can't consider such a number all at once, or survey it altogether. Now that you've finished counting, you can appreciate the magnitudes represented by a thousand years or a thousand souls. And that a thousand times a thousand is utterly outside your range of visualization, while a billion (a thousand times a thousand times a thousand) may be comprehensible to the reason, but it remains a blank to the emotions. It is unknowably many. And now, just for an instant, think of humankind today.

23 *Dread the arrival of the bus*

Duration: *5 to 10 minutes*

Props: *a bus service*

Effect: *relief*

The act of waiting has two aspects. An opportunity for a bit of peace and contemplation on the one hand: you have nothing to do but wait until the expected moment arrives. This kind of passivity can actually be a real source of pleasure. Time elapses, whatever happens, and that certainty has something reassuring about it. But waiting can also terrify–what will transpire is by definition never entirely controllable or predictable. Explore this kind of fear, which has no explicit pretext; extend it, magnify it as if under a glass, increase its intensity and its duration.

Take your place at a bus stop. Very often there's a moment of slack before the bus arrives. You don't know exactly how much time you'll have to wait. The bus has got stuck in a jam. Or it has broken down. Demonstrating crowds have held it up for ages. You're going to be late. You must find an alternative means of transportation, provide explanations, telephone people to warn them, even change your timetable. The entire day may be put out, every meeting will have to be pushed back. Picture in your mind a long series of chaotic outcomes.

Start with a minor and banal anxiety such as this. Transpose and embellish. Maybe the bus will arrive, but driven by terrorists, stuffed with dynamite, and with no brakes left. It will trigger a series of uncontainable catastrophes: there is a new virus on board, some murderous bacteriological weapon. The driver is an extraterrestrial, and the passengers are all in league with him. All those who got on at the preceding stops have already met their deaths with bloodcurdling screams.

Continue, amplify, exaggerate. It doesn't matter if it all seems grotesque, that you don't believe a word of it, that you can laugh it all off, remaining convinced that a perfectly ordinary bus will be arriving shortly. The important thing is that you should experiment with apprehension of this kind, however minor its cause, and with moments when certainty becomes slightly unfixed. The very fact of having invented these fantastic hypotheses will leave a trace. The vague idea of some dreadful event remains there like a tiny crack, a potential weakness in the normal course of affairs.

Here's the bus. You get on. Everything seems to be as usual . . . But are you so sure?

24 Run in a graveyard

Duration: *1 hour*

Props: *running shoes, large cemetery*

Effect: *pious*

Graveyards: peaceful and pacifying enclosures. Spaces suitable for meditation, and any other kind of reverie as well. With flowers, and without people—a double attraction. A few mourners, a few gardeners. The odd tourist walking by, a graveyard enthusiast reading the inscriptions.

The idea of going for a good long run in such a place may seem rather shocking. An inappropriate provocation, a silly prank. It may even be a misdemeanor, which will earn you a rebuke and a fine, payable to the municipal court. Or else it is an offense—to the grief of families, and to the memory of the dead, according to some unwritten but universal consensus. The idea may seem unacceptable for other reasons, perhaps more profound and less easily definable ones: it offends against the order of things, the way the living behave toward the dead. The watchful immobility of the living compared with the utter stasis of the dear departed. The former breathe, the latter don't. It doesn't do to accentuate the contrast. In the very spot where those who were once alive now repose, without words or motion, shouting and gesticulating has no place. The runner-between-graves lays himself open to summary justice.

Don't let yourself be scared off. Negotiate the obstacle and overcome the shame. Meaning will come in time, as usual. Deal with the practicalities first of all: wear a tough pair of shoes (the alleys are frequently stony and uneven), and choose a graveyard that is big enough. Most country graveyards, while making pleasant places to walk among family tombs, are useless for long-distance running.

So here you are at last, embarked on this strange experiment. To begin with you'll feel, naturally enough, some remnants of embarrassment, the feeling that you're acting in an incongruous and unseemly fashion. You think of all the skeletons lying in their coffins, stacked on top of each other, piled up, shrunken, damp, dark, almost all of them forgotten. And you find that your fleetness of foot is out of place. Moving like that, in such a lively way, among the petrified—this is conduct unbecoming.

It may be helpful to concentrate on this discrepancy, and enjoy it. You, after all, are alive, able to run, and rejoice in the power of movement. Not they. Well, too bad. And so much the better for you. A beating heart, hot blood in your veins. They know nothing of all that now, they have left life and time behind. But you, you are moving through the soft air, with your feet arching over the earth.

The experiment is interesting only if you can get beyond this first stage. Try now to dissolve this aloofness that sets you rejoicing apart from them. You start to feel that at the very heart of your progress you are motionless.

That there's no separation, in the end, between movement and stillness. Your strides may well be long and regular, your stamina considerable, immobility seeps into everything. What you come to feel now is the presence of stillness at the heart of movement, of repose at the heart of the race. And of respect within transgression. You are no disturber of the dead. Running between their graves, unmindful of their names as of the proprieties, you love them.

25 *Play the fool*

Duration:	*30 to 40 years*
Props:	*a complex society*
Effect:	*joyful*

How did the court jesters amuse themselves, when there were jesters and people had a riotous good time? They made fun of one and all, and had no time for rules and conventions. They talked too loudly and laughed at all the wrong things. Defying expectation was their destiny. They could shake up people and conventions. Nomadic and subversive, they wandered roads and crossed rivers, weaving in and out between etiquette and obligation. They overturned holy images, parodied the sacraments, and mocked the authority of the Church.

Let us do likewise. Not that the rivers and roads used by those vociferous singing gangs are still at our disposal. Try setting out like that today and you will be rapidly locked up. So find another way. Make yourself into your own critic, journalist, writer, novelist, filmmaker, musician, drummer—something of the kind. Something out of step. Do everything you can to make waves. Don't dream of changing the course of history, just sow a little chaos around you. Disorganize plans, create surprises, confound predictions. Live stubbornly within your society, without at heart acquiescing to it.

Obviously you have to submit to certain norms and forces. You might even have to crawl, out of prudence, cowardice, or even sheer cheek, in front of some Mr. Big or other. Tell yourself that is unimportant. Give way, kowtow tactically, from time to time, if you are absolutely certain that something within you is resolute and unbending.

Take care to leave yourself room to maneuver, over a long period. Act slantwise. Move like the bishop in chess— systematically diagonal. Walk crablike and crossways. Day in, day out, meeting no resistance. Make it a habit to seek the least appropriate, most incongruous answer to any question. Apply it from time to time, and see what happens.

The longest and hardest thing about playing the fool is arriving at the realization that truly nothing is serious. Occupy the horizon, that point of convergence where absolutely everything becomes, in a sense, laughable: existence, death, humanity, love, the universe, ants, writing, money, careers, bodies, thought, politics. Among other things. Not forgetting laughter itself, and hilarity, and court jesters.

26 Watch a woman at her window

Duration: *a few seconds*

Props: *chance*

Effect: *reverie*

She is dreaming, or washing the dishes. She's looking out at the street, or smoking. Or hanging out the washing. She's twenty or sixty, poor or rich, beautiful or otherwise. She watches you, and your eyes meet. Or else you see nothing but her profile and her abstracted gaze. You, too, enter upon the scene from a variety of directions. You might be looking down on her from above, or from the same height, or from below, as you pass by on the street. You see her face clearly, or just her torso, because her head is bowed, or partially concealed by a curtain. You can clearly make out her clothing, the curve of her breasts, the fall of her shoulder, the rounded flesh of her arm. Or else you see no more than a vague silhouette, scarcely more than a shadow. Whatever.

In every case you feel the same emotion. You are contemplating an unknown woman, in the privacy of her home, who is only partially and fleetingly visible. Her presence is framed in the window. Naturally, you know nothing about her. You are just passing through. It's highly unlikely that anything will happen between you. She will never be anything more than a dream and a brief illusion. Nothing else. And you know it.

But there's nothing to stop you from fantasizing. She's seen you, she signals to you, you will join her and embark upon an impossible affair, a sweet and terrible passion, unforeseeable, providential, made up as much of obscenity as of tenderness. And from the first second, she too will know, as one knows those impossible things that we don't understand yet prove decisive, so that everything is changed forever.

A male is writing these lines. For the female reader, the experiment must be adapted, or reinvented. I wouldn't presume to influence, even were it possible to do so, such a personal scenario in which desire and the gaze, sex and its description, are so intermingled.

You've reached the end of the street. Nothing has happened. Life goes on. Until next time, then.

27 Invent lives for yourself

Duration: *a few months*

Props: *none*

Effect: *disturbing*

You only live once, as they say here. Others, elsewhere, will assure you that you have already lived out several previous lives. No matter. You can multiply your own lives yourself, and feel them proliferating. To do so you need to carry out a relatively long and fairly demanding experiment, but whose effects are well worth the effort.

During a period of several weeks, try systematically inventing lives for yourself. Tell your new barber you were a taxi driver in Detroit before you delivered pizzas in New York. Recount your years of teaching in Australia to a distant cousin. To your nephews spin yarns about places you never saw, livelihoods you never made (who's to know, after all?), great and small adventures, of fox hunts and fogbound ports.

Do it properly. Don't just tinker. Recount the same stories several times. Spice up the anecdotes, add new details, fill in the blanks and eliminate implausibilities. Tell the same stories to the same people. Take care not to get muddled up. If need be, take notes, fill out cards, do research. Persevere.

After a few months, you'll be familiar with these alternative lives. You'll have answered a lot of questions, and explained a good deal. You will have described, narrated, taken up, and repeated the key episodes of your various parallel autobiographies. Above all you'll have implanted your fabrications into the minds of people who believe what you've told them, and who will pass them on to others in the version you made up for them. They believe it.

Why don't you? The point you need to reach is when you start to doubt whether it's all false, and when you can't quite tell what belongs to fiction, and what to your real life. Or when—it comes to the same thing—you can admit to yourself (without forcing or sudden delirium) that what you used to consider your "true life" is really, in fact, just one fiction among others. No more, no less.

28 Look at people from a moving car

Duration: *10 to 40 minutes*

Props: *car, driver, big city*

Effect: *invisible man*

To try this experiment you need to be a passenger. When you are the driver your attention is necessarily too taken up by steering and safety to allow yourself to be absorbed by the surroundings. But when you are being driven, there is nothing to stop you from giving yourself over to the pleasures of secret observation. Passive, abstracted, floating. Moving without having to move a limb, cruising at greater or lesser speed, watching without being watched, the car passenger is actually perfectly placed to be a voyeur.

So climb into the back of a taxi, or a friend's car (the context doesn't really matter), and flow with the movement. It works best in a big city. Decide that you're a kind of flying carpet or in a box floating along at ground level, from where you can see the world pass by without it seeing you. You pass among people and they don't see you. You notice some moment of their lives—interrogative, anxious, irritated, desiring, anticipating. Clothes, postures, bottoms, shoulders, differences of age, hairstyles, fashions, wearinesses–it all files past. And then suddenly, in the middle of this maelstrom, is an unforgettable gaze. Or the perfect profile of a perfect stranger. Instantaneous emotions, two-a-penny eternal instants instantly forgotten.

Between yourself and these living beings–these existences fraught with drama, anxiety, and countless projects–there is movement and a window. You can always open the window, but you'll still be just passing through. Do the experiment often, and in several places, where possible changing country and continent. Draw any conclusion that you wish. There are several of them.

29

Follow the movement of ants

Duration: *about 30 minutes*

Props: *an anthill*

Effect: *reflective*

We've all done it, but it still works. To spend a bit of time observing a column of ants is thought-provoking. Attend to their dogged regularity. And note, even if you've done so a hundred times before, how they follow each other, pass each other, construct the regularly moving thread of their passage. Grasp the general design of their journeys, with their tiny individual variations and momentary turnings-back. Spot the weary heroisms and the implausible ferryings.

And then take up those banal reflections everyone has already made. Ask yourself how such a life is to be conceived. Ponder the idea of a biological community, of a society without language. Perplex yourself with imagining an inhuman city. Gulp in front of the depiction of an organism made up of a multiplicity of individuals. Reread your *Micromégas,* your Fabre, your *Bouvard et Pécuchet.*

Finally, try to imagine yourself as an ant. Pushing a breadcrumb, crossing a pebble, moving around a piece of broken glass. How do you know where you're going? What tasks you're meant to be accomplishing? Are you hungry? What does that mean? What are you thinking about? And what does that mean? What's it like—being an ant?

You know these questions have no answers. There do exist parallel worlds, opaque to each other, non-communicating, and we are really in error when we speak of a single universe. Planet Ant is not Human Earth. It's not included within it, hardly situated within it. You will conclude that the plurality of worlds exists under our nose, is always there, and we still don't really under-stand the first thing about it, the why and wherefore.

In short you will find that the ants provoke in us no new ideas, and certainly no interesting ones.

30

Eat a nameless substance

Duration: *a few minutes*

Props: *something edible and anonymous*

Effect: *troubling*

As usual, something banal to begin with. The only thing that changes it is the act of attention, of spending time. Quarry the instant, delve into what happens during the moment otherwise apparently without interest. Bore into an action or a feeling. Go down as far as you can. Sometimes one has to give up, having found nothing worthwhile, encountering simply a dead end or an impasse. But at other times, a turning in an underground tunnel suddenly gives onto a gulf, a cavern, a vast and somber grotto spreading out below the surface.

Example: you have often eaten a food whose name you are unaware of. You probably didn't pay that much attention. The circumstances can vary: a foreign country whose customs and language are unknown to you, a regional exoticism, a local speciality, a meal with friends, a visit to a grocery that sells exotic produce. In short, you have eaten something you can't give a name to, so you cannot say to a friend: "I ate some — or some —." You have to use a long periphrasis, describing its color, texture, smell, and taste by means of a series of comparisons and combinations. "It's like — but less — and more —, it smells of — but with the taste of —, and the color of —."

The next time this happens, at the moment you taste this nameless "thing," stop. Analyze exactly what is happening. No matter if it tastes nice or nasty, though it's preferable if you like it and are thus prompted to ask: "What is it I lack, exactly, when I lack the name for something? There's nothing wrong with the taste, which is very idiosyncratic, and there's nothing wrong with any other aspect of this food." And yet, the fact of not knowing its name makes the food seem somehow abnormal. Incomplete, incongruous, unclassifiable—for as long as you don't know its name.

As soon as you learn the name the situation changes. You'll still like or dislike it—that remains the same. You'll recognize all its former qualities, but from a different angle. The substance will henceforth have entered that network of elements that words master; it will take its place within the chain of classification.

It would be too much to say that knowing the name changes the taste. But it definitely does alter our attitude toward the taste, our way of considering it. We taste the unknown food more interrogatively, quizzically, exploratively. By contrast, once we know the name, we are eating the noun, we are ingesting layers of language, digesting slices of phoneme.

And then the suspicion may arise that what we really eat are words rather than foods. Our appetite is as much linguistic as gastric. The tongue that tastes is not just in our mouths. It is in our dictionaries.

31 Watch dust in the sun

Duration:	*15 to 30 minutes*
Props:	*a room, a ray of sunlight*
Effect:	*reassuring*

A room that is fairly dark. The shutters almost closed. A ray of light pierces through the blinds. A strong slantlight, the rays of dawn or sunset. The light piercing the gloom is filled with tiny glittering bodies. This must count as one of the most moving and magical spectacles it is given humans to see. Spiraling, turning, crossing, and recrossing, thousands of minuscule sparks that hold and refract the brilliance. Dots, tips, microscopic feathers, tiny flecks, minute airborne objects, light and dancing, that pass through the light in a way that is sublime, grave, joyful; fearfully busy, moving in whirls and arcs impossible to follow, in fragmentary trajectories, pure glints of existence.

The most exquisite thing about this miracle of scintillation is its density. Leave to one side, if they come to you, your memories of childhood, of old-fashioned games, of houses in the country, the smell of cupboards. Concentrate wholly on these astonishing specks. The border between light and dark is suddenly so emphatic, clear, and direct that you feel you could almost touch it. The teeming bodies pass in and out, either side of the frontier. This is the space to dream in.

Few simple experiments give us such a strong feeling of an invisible world suddenly revealed. There in the ray of light, like a slice of different space inserted into our own, is a universe from the other side, from inside out, from elsewhere. Suddenly made visible, as if by breaking and entering. What would the world be like if we could see dust glinting all of the time, everywhere, unceasingly? Does there not exist, everywhere and unceasingly, a stratum of existence that is both invisible and present? A plane we might be able to reach, a different space, contained within the one we know?

And what if, to reach it, all we needed was to know how to adjust the shutters correctly?

32 *Resist tiredness*

Duration: *variable, several hours*

Props: *none*

Effect: *subtle*

Life and tiredness are inseparable. It is useless to dream of a pure, blissful rest, which effaces both effort and tension. Life is constantly prey to the depletion of energy, to lassitude, and to real or imagined aches and pains. To persevere brings on fatigue, if by that one means the result of effort, not of depression. Too many people complain of a melancholy lassitude or an oppressive inevitability that crushes and drowns them. We succumb when we can fight no more, finally conquered by our lack of strength, comprehensively and utterly beaten.

A little more of this feeling, and tiredness becomes an ocean in which we perish without trace, a quagmire that swallows us up for good. The seriously exhausted are soon reported missing. They sink, slowly but surely. Country people know the signs: when they say someone is "tired," they mean he'll die very shortly.

So we must struggle against this notion of inevitable decline, irreversible exhaustion. We should, first of all, rid ourselves of the idea that there is only one type of fatigue.

We should distinguish between the different types, and realize that they have almost nothing in common. One of the surest ways of resisting tiredness is not, in fact, defying it by gathering one's last forces to confront the tide. We should, on the contrary, go with our tiredness. Not resist it head-on, but learn to ride its currents. We should not consider it as an obstacle but as a means of going forward–tiredness should be a vehicle, an instrument to steer by.

Try to cross from one type of tiredness to another. Get used to opposing them. Try to ascertain the type of tiredness that suits you best, and those you should beware of most. To this end, practice walking in the heat, not sleeping, working excessively and without a break. Try doing several things at once, make love more and more frequently, and differently, try to do everything all the time.

Whatever happens, you'll reach an end. The question remaining is to know whether that last realization reassures or worries you.

33 Overeat

Duration: *2 to 3 hours*

Props: *copious quantities of food*

Effect: *dyspeptic*

The context matters little. A meal with friends, a social obligation, a family reunion, a gastronomic tour of some rich and abundant part of France . . . all sorts of occasions lend themselves, in our opulent countries, to this banal conclusion: you have overeaten. You're not really comfortable. Your stomach is overfull, your head heavy, your mind blurred, and your mouth furred up. Among the supplementary discomforts we might list headache, palpitations, hot sweats, hot and cold fevers, regurgitation, and gas.

The experiment consists in transforming this chance eventuality into a learning curve. Begin by putting to one side all considerations relating to your resolutions, and every spasm of resentment against the people who have landed you in this situation. Don't be angry with yourself or others. Face things squarely, as they are: you have eaten too much, more than your body can cope with without discomfort. It's a fact. And now, go with it.

Follow without resistance or reflection the numerous fluctuations you are passing through: torpor, wakefulness, doziness, lucidity, heaviness, relative lightness. Attend to the slow struggle that has been engaged, inside your digestive tract, between this mass of ingurgitated food and the reconstitution of your body's metabolic balance. Once again: rather than suffering this discomfort as an unpleasant inevitability, turn it into the starting point of an exploration, albeit a somewhat crude one, of your relation to reality.

Examine, for example, how your consciousness is modified by a stew, the particular type of torpor linked to a heavy cassoulet, the hot flushes triggered by fried *foie gras*. The point of this is not to classify recipes according to their effect, but rather to follow the metamorphoses of your own identity along the gastric passages.

Clearly you are not "the same" at any given moment in these circumstances. Just ask what becomes of us, with our fine speeches on free will, consciousness, personality, reason, the moral law, and other vast questions, when some feculent or other can change the complexion of our universe, and a bit of fat can bring us low. Our sharpness of intellect gives up after a few platefuls. Something to bear in mind.

34 Play the animal

Duration: *10 to 20 minutes (to be repeated)*
Props: *none*
Effect: *shapeshifting*

Shut the door. For the duration of this experiment it is vital you should not be disturbed. The moment you are sure of being alone somewhere quiet, start imitating the animal you like the best. For instance, pant like a dog, tongue hanging out, with a throaty noise at each breath. Sniff the carpet noisily, and especially the furniture legs, turn in circles, lay your cheek against the ground, gnaw on your forearm or your elbow, etc. Or else, depending on your mood and your talents, meow, cluck, roar, hiss, neigh, bellow, bray. And perform the accompanying actions.

Do not imitate! The point is not to reproduce sounds or postures. Your talent for farmyard noises is no help here. In fact it is a hindrance. What you must do is get "under the skin" of your chosen animal. Act it out as it comes naturally, with minimal or no control. When you feel the need, start groaning or moaning. Disrupt the rhythm of your breathing, crawl along the ground. Rub your head against the wall or the floor. Dribble, lick yourself, adopt clumsy or fastidious movements. If need be, undergo a change of teeth, muscles, smell. Acquire claws, a beak, plumage, horns, as and when you need them. Delve into yourself to find more of these strange paths. You should repeat the experiment and take it further. No result is guaranteed. Nothing to be understood. Everything to be tried.

One thing you will discover very quickly is that some routes are passable, and others not. It's relatively easy to be a wolf, a lion, an elephant, or a hyena, an antelope, a polar bear. But once you leave the mammal family things get much harder, with the odd exception or unless you're a natural adept. Becoming an ant, a tick, a flea, or a spider is in each case very difficult. Snakes, worms, and invertebrates in general are not easily accessible. The huge families of fish, birds, and mollusks are relatively intractable to the amateur shapeshifter. Not to speak of the vast empire of bacteria, which remains almost entirely closed.

The world is small, after all.

35 Contemplate a dead bird

Duration:	*10 to 15 minutes*
Props:	*a dead bird, preferably dead for several days*
Effect:	*meditative*

They're to be found pretty much everywhere in the country. Especially in spring or in the middle of summer. If you go walking a lot, you're bound to find one. A fledgling fallen from the nest, or a juvenile attacked by a hawk, or an adult riddled with shot. The causes are not your concern. The why and the wherefore. Rather than walk on, leaving the dead bird behind, you will stop and contemplate it.

Look closely at the dulled feathers, often covered with dust or a bit of earth. Observe its eye, colored or whitish or eaten out, and the ants coming and going, and maybe a few maggots. Note its claws, quite still, abandoned, twisted. Look for the bones, so slender and so visible. Above all remark the whole attitude of abasement and loss, the way a dead bird is so thoroughly a corpse, muddied and humiliated, in the truest sense, and yet how it knows nothing of all this, and escapes from it into a depth quite alien to sleep.

If you look closely enough, you'll probably find the sight a sad one initially. A life snuffed out. A body misplaced, a bird lying on the ground, all stiff. Something resembling defeat and failure. The experiment consists in going beyond that, by seeing more and more clearly and distinctly.

You see that the bird will never live again. And also that it feels nothing. That this is how it is, beyond help and complaint. Innocent of nostalgia or recrimination. The longer you look the clearer it should become that there is nothing, concerning this little corpse, that can be cause for regret. There is only the present. And you start to realize that it is perfect. Because it is the only tense there is.

At first, this is incomprehensible. Strictly speaking, it may never be given us really to understand, only to feel. What you will grasp, however, if you open your eyes wide enough, is that there is no other world to see. That everything, absolutely everything, is here and now. In the present, as it occurs. There is nothing elsewhere, or before, or anywhere in space or time, that is different, better, preferable, comparable, regrettable. Nothing but this.

36 Come across a childhood toy

Duration: *unpredictable*

Props: *a toy that belonged to you*

Effect: *multiplication*

You are emptying an attic or a cellar. Inheritance, grand-parents, country uncle. Or just an old trunk full of personal relics. Or, less likely, you come upon it by chance in a junk shop. Whatever the circumstances, a forgotten toy comes to light. It has to have been forgotten. Totally. You have no recollection of it. You carry in your head a few faded images of objects once familiar to you during your earliest years. But not this one. This toy had disappeared from your memory. You would never have been able to bring it to mind, or look for it.

But you recognize it the moment it appears. Without hesitating, you remember every aspect of it. You know this toy; it is utterly familiar, part of your everyday existence. It is yours. You rediscover that same paintwork, each scratch, each pencil mark. A tiny crack, a rough edge, the little bit that's missing, you know it by heart as something obvious, a given. And you are suddenly transported into the world of this toy, into its own time and its own particular space. You are riveted by its presence, but without being snatched from your present reality.

How is this possible? How is it that these vivid, numerous, and precise details can be both inaccessible and yet perpetually available? Vanished, but not effaced. Instantaneously recuperated, or rather recuperable. Why, exactly? Could there be worlds contained in the present, enfolded within it, without us knowing? Could it be that we move forward, surrounded by unused, available lives and virtual existences?

37 *Wait while doing nothing*

Duration: *10 minutes to several hours*

Props: *waiting room or similar*

Effect: *calming*

This is a particular kind of waiting: you can do nothing about it because you know what the outcome will be, yet you don't know how long it will take. The doctor's waiting room, a government office, an airport or a train station—especially during a delay—are suitable places for this experiment. You know that the doctor will see you in the end, that you'll be called to the window, that the plane will take off and the train will eventually arrive at the station. You can see it's a very different kind of waiting from that in which the outcome is uncertain, and possibly disturbing. Moreover, you are forced to be passive: there's nothing you can do to speed up the process. You are squarely confronted with duration, with the unavoidable passage of time, passing more or less slowly, more or less viscous.

A lot of people find this situation difficult to put up with. They contrive to avoid this head-on encounter with time by reading magazines, novels, essays, taking notes, consulting a filofax, using a cell phone, working on a laptop, or simply watching the world go by. In short, they keep busy, filling this given span of time with activities, with big or little ideas, with a variety of tasks.

You ought to experiment with exactly the reverse. Do nothing. Without becoming either irritated or bored. Let yourself float in time, knowing that it will pass, inexorably, in you and without you. You should merge without anxiety into this total passivity. Everything will happen, and nothing depends on you. You can be empty, amorphous, immobile, indifferent, dreamy, absent—time moves on regardless, and this interval will come to an end. You can thus make the discovery that there's no need to kill time. It dies by itself, on its own, unceasingly.

38 *Try not to think*

Duration: *10, then 20, then 30 minutes*

Props: *none*

Effect: *void*

This is an experiment that takes us to the limits. Not to think at all, when one is wide awake and in full possession of one's faculties, cannot be achieved, or only for very brief intervals. So it can only be attempted. Some attempts are short-lived, some go further. Some come close, other less so, to the impossible goal. Some brush it, others only glimpse it on the horizon.

Why is not-thinking impossible? The experience of it would remove us from the sphere of the human, it would allow us to escape the incessant babble of language. We would tumble into a state of stupefaction, into pure, moment-to-moment, animal life. Or else, which is possibly the same thing, we would fall into the divine, the bottomless, abyssal silence. It may be that thought is a patchwork thing existing in-between. Not quite divine, and not quite stupefaction. It may be a way of rowing between eternity and the instant. Or between silence and words, presence and absence, being and non-being, etc.

In any case, thinking cannot be arrested definitively. It's more a matter of temporary interruptions, circumscribed parentheses. These are possible, and worth experimenting with. To launch into this, you must take it little by little, in measured stages. It is vital, first of all, that you don't tense up, that you let yourself go. Willpower, here, can only act obliquely and indirectly. What we're dealing with is not an achievable project, for it's obviously of no help to think that we are not thinking. It's better to know in advance that we are going to fail. That we shall be, at one moment or another, caught thinking. Failure is certain. Therefore any progress is of value.

The most effective training consists in letting your thoughts flow by. Don't stop them (impossible) but don't hold on to them (possible). Observe them as you do passing clouds, far off and inevitable. Imitate the indifference of the sky. Persevere in remaining yourself unclouded, and pay no attention to what is passing by. Remain at one remove, somewhere below the frame, your eyes open on what is in front of you. And that is all. Sensations still exist (colors, light, breath, your skin, your muscles, noises off) but don't integrate them into your consciousness, still less into an idea or an argument. And finally, occasionally, in snatches, you may manage to move forward into the clear sky, into the empty light, where there is no disturbance, and no form.

These brief successes can have substantial consequences. Repercussions that go way beyond the moments during which they occur. Even one such success will have a lasting effect.

39 Go to the hairdresser

Duration: *about 1 hour*

Props: *a hairdresser's shop*

Effect: *disheveling*

It appears straightforward. You go in, you're given a shampoo, and then someone cuts your hair—a little, a lot, or none at all. The experiment consists first of all in feeling the extent to which this apparently banal situation is actually much more complicated than it first appears. You might imagine that your hair is not in fact an entirely indifferent part of your body. It's rather hard to gauge its exact relation to your body. Is it dead, or alive? Without feeling, or traversed by a different kind of nerve-system? Is it outside your body, or inside it? Between the two?

What happens when it is cut? It may be that your hair is linked directly to your ideas, and you may never have the same ideas again after you leave the hairdresser's. Your soul will have a new haircut, it will be unrecognizable and unusable—you'll no longer be at home in yourself. You'll be something completely other, disorganized from within.

Or else your hair appointment changes your whole appearance. You'll leave with a different head, with everything different, the shape of your nose and the color of your eyes, the roundness of your cheeks. Your whole body will be altered, you'll be taller, or smaller, stunted, or swollen.

Or perhaps the hairdressers are archangels, messengers of God, neighborly redeemers. You'll leave the salon transfigured, in a glorified body, borne aloft on the music of the spheres and in the joy of salvation.

Lost or saved, you are about to live through a decisive moment. You have a rendezvous with your own destiny. An unparalleled transformation is about to take place. Through some dreadful operation, some unspeakable alchemy, your hair appointment will leave you breathless and broken, the victim of internal cataclysms. This is what you have come to believe.

The second part of the experiment consists in washing away these fantasies with a medicinal shampoo. You know full well that nothing is going to happen. You are going for a haircut, that's all. It'll be a good one, or it won't, it'll conform more or less exactly to what you had in mind, to your "aesthetic" notion of what suits you. But these minor variations are without consequence. They are without the slightest importance.

Your dreams will have been for nothing. But you will at least have experienced the distance that separates your phantasmagoria from reality. Reality is nearly always flat, banal, straightforward, without contour. Reassuring, in a way.

40

Shower with your eyes closed

Duration: *5 to 10 minutes*

Props: *a shower*

Effect: *peaceful*

You do not know where the water's coming from. You can imagine, with your eyes closed, something other than a shower: tropical rain, for example. Perhaps you are not alone, there are other people watching. A scene starts to take shape, possibly a whole story. Close it down. That is not what this is about.

Just feel the warm water, the drops, the jets, the pressure. Feel only that, see nothing, and hear nothing, other than the sound of the water. Try to confine yourself to this single sensation, without meaning and without images, there under the water. Stay like this, standing under the water, with your face held up toward the shower. Don't look for words to describe it. Remain speechless, and gather your whole self within the weight of the water. Nothing but the sensation, or as good as.

Do not melt or merge, just remain uninterruptedly drenched. Avoid anything that smacks of prayer, ecstasy, shamanic journey. Just remain there, with your skin wet. Thoughts come, ideas arise. A few stubborn questions will not go away. Anxieties likewise. The water passes over you, cleanses you. Do not step out of that shower jet's narrow circle.

41 *Sleep on your front in the sun*

Duration: *about 1 hour*

Props: *bathing towel, sun*

Effect: *defiant*

Initially you feel just a light torpor, a drowsiness. Then the pressure of your chest against the ground becomes greater, your breathing slows, shapes grow blurred, cries in the distance get fainter. All you feel is the heat stretched upon your back like a comforting blanket, and a dryness on your legs. A light breeze is just enough to moderate the intense heat. All you feel is the protection given by the hot light, its uninterrupted covering. You're entirely enveloped in its gentleness, naked, relaxed, and confident you won't get cold. This thought alone comforts you as you start drowsing off. Which you do.

A little while later the experience of waking is strange and not exactly pleasant. You are not sure where you are, why you fell asleep, and how to emerge from it. You feel the great heat that has invaded your skin. You start worrying about burning. And the cancers you've exposed yourself to, the hidden melanomas, the spreading metastases. As you wait for your coffee, you start to experience a rather furtive kind of agony.

Don't go with it. Say Up Yours to medicine, to prudence, to the merchants of fear and wisdom. Refuse to live in an air-conditioned cell where you will feed on nothing but endives and mushrooms, drink nothing but spring water, and be protected from all outside radiation. Only remember that you are going to die anyway, sooner or later, more or less horribly. In the meantime, sleep where and when you please. And the same goes for the rest.

42

Go to the circus

Duration: *2 to 3 hours*

Props: *a circus*

Effect: *humanizing*

Beware of people who do not like the circus. They are undoubtedly too efficient and too sure of themselves—ruthless. To understand the circus, even if you're not particularly attracted to it, experiment with sitting near the circle. Choose a small circus, preferably, nothing too splendid, rather somewhat impoverished. Avoid Madison Square Garden, Barnum, and the other big concerns. With them, it's harder to grasp what makes the circus so moving—its mixture of misery and reverie.

For usually these places have something sordid about them—which is both intrinsic and necessary. The sawdust on the track, the smell of animal dung, the dust from old marquees, the whiff of sweat below the tent canvas. It must also be a closed space: the circular ring, the canvas heaven, the guardrails. The circus encloses a space proper to itself, a world not to be confused with the rest of the universe. You can define the circus rather as you can the human world itself.

In this circumscribed sphere, a bubble of dreams is constructed. In a very elementary or even stupid or vulgar fashion: sequins and paste and all that glitters. Heavy fake jewelry. False luxury, false chic, a factitious facility and a forced gaiety against a background of grinding sadness. This is what makes the circus so moving, an exemplary model of the human: doggedly constructing laughable dreams out of the filth and the muck. Every evening at 8:30, with a Sunday matinée at 3 o'clock.

You should head toward the circus tent. Line up for a bit, and pay too much for the discomfort, the staleness and bad smells. For the uncomfortable seat. You will easily surmount these drawbacks, and by watching the lightness of the acrobats and the skill of the conjurors you will feel you have escaped the crushing sense of failure. You'll even start to dream of a humanity full of crystal balls, lit by spotlights, smiling into the brass band, happy amid the cotton candy. The performers on stage will come to seem almost beautiful, courageous, worthy, full of virtues, capable of grand exploits, larger than life, their bodies shining like those of gods, so supple they are, and light, swift, and aerial. For a while float within this glittering bubble.

And then, crucially and most moving of all, something goes wrong. A juggler drops a ball, a trapeze misses, one of the animals remains obstinately motionless. You notice that the beautiful contortionist has a hole in her tights. Brusquely, you see something pitiful, some pride brought down to earth. Some terrestrial dream, smudged as it always is, and always more or less wounded. A shattering failure. An image of human doggedness. You should go back to the circus time and again.

Try on clothes

Duration:	*30 to 50 minutes*
Props:	*ready-to-wear shop*
Effect:	*reverie*

For thousands of years now clothes have done more than just protect us from the cold or the rain, or preserve some supposed modesty. It is hard to imagine that even the most primitive garments served a merely thermal function. No doubt they had a symbolic role. It's worth noting in passing that not one of the societies revealed to us by anthropologists used clothes for a single, merely practical end. Clothing is always coded, involved in power games, norms, and social role-playing.

We have multiplied a thousandfold appearances and meanings. Clothes indicate the social and physical milieu we come from, the specific powers we exert or the domination we undergo; they signal class, character, age, job, sex life, transgression, submission. They can say: "I'm a youth from the suburbs seeking to escape humiliation by wearing the same brand-names as the bourgeois kids my age, but I choose different colors and mix them up in a way that seems to them out of place, without realizing what I've done." Or else: "I'm a rich bourgeois from the smart end of town, my children have grown up, my husband bores me, my lover likewise, but you can try your luck if you know the code and the right way of summoning the maître d'hôtel."

You can experiment with trying on clothes, not to buy them, but to explore unlikely styles and looks. Instead of searching as you usually do for something that suits you, that matches your taste, your status, and your size, your morphology and your fantasy life, try on some incongruous apparel. Too young or too old for you, too smart or too vulgar, too loud or too sober. Rags that seem in any case inadequate, excessive, out of synch. And that make you smile each time you see yourself thus garbed.

Imagine you are one of those cardboard cut-outs in a children's game, that are dressed in all sorts of different ways by means of those little paper tongues that fold back behind the shoulders. Dream of yourself as Barbie or Ken. Do your best to come on as a rocker, a diplomat, a salesman, a rapper, a peasant, a butcher, a designer, a duck stalker, an intellectual, a janitor, an athlete, a junior executive. And every time, compose in your head the life that goes with these fabrics: speech patterns, eating habits, house and home, leisure activities, holidays. Then put everything back on a hanger. Thank the salesperson.

44 Calligraphize

Duration: *20 to 30 minutes*
Props: *paper, a good pen*
Effect: *concentrating*

Writing is not an intellectual activity. It is first of all a manual exercise. Preeminently so, perhaps. One's thoughts, and the supposed or imagined communications one inscribes on the paper are arguably far less important and of less interest than the attention required to form letters beautifully, tracing their contours precisely and aesthetically, playing with the tiny balances of curves and uprights, loops and dots.

To try this out, you should begin by writing out the first sentences that go through your head, however banal, but in doing so write with a perfectly regular rhythm, without pausing. Once again, what matters is not the content of what you are writing—that is of secondary importance. The only thing worthwhile is the regularity of the characters, their measured forward movement and the fact of inscribing, in order, joined-up letters that are well formed, legible, proportionate, distinct.

Concentrate your attention on the precise and tiny movements of your muscles and the brief trajectories of your pencil or your pen. Try as far as possible to avoid stopping between two sentences. Keep up the same rhythm. What you write doesn't matter. The act of writing is enough. Try not to speed up or slow down—your script should be polished and clear but equally, in its forward movement, it should be steady, monotonous, and fluid. Try to attain a constant speed, an almost perfect automatic continuity. Once again, the only thing that counts here is the fact of your covering the paper, inexorably and with an application less and less dependent on your willpower, with horizontal lines made up of an obstinate succession of letters and words.

You can write about anything that occurs to you—childhood memories, shopping lists, novel insults, a parody police report, holiday postcards, intimate confessions, a love letter, a tax return, an eyewitness account of an accident. Whatever you choose, it is vital that you worry as little as possible about what the sentences mean. Whatever your sentences say, you should consider them merely as vehicles enabling the forward movement of your script.

The experiment consists in sensing how the forward movement of writing, the calligraphy covering page after page, is quite indifferent to what the sentences mean. On the one hand there's the whole maelstrom of ideas, of syntax and sentiment, of meanings that teem and proliferate, of harmonies and discords. On the other hand (but does an "on the other hand" really exist?), we have the regular pulse of script, fantastic, almost pure, automatic, secreted by nothing other than the necessity of going forward, by the repeated regularity of being incessantly identical to itself.

You may then come to feel that everything we believe we are saying or thinking is double. Over and beyond the transmission of an achieved meaning, at once already known and permanently available, you may get a glimpse of the secret permanence, the inexhaustible source–impossible to discern–at the heart of this flowing script carried forward by its own momentum. It has nothing to do with what the words mean. Is completely indifferent to what texts transmit in terms of information, ideas, feelings. Script, nothing but script. Traversing bodies, thoughts, muscles, sheets of paper. The interminable flux of writing.

45 Light a fire in the hearth

Duration: *15 to 20 minutes*

Props: *fireplace, wood, newspaper*

Effect: *primitive*

Let us suppose that laying a fire in the fireplace is a very ancient ritual whose meaning you have lost. You know a thing or two about it, however, a few tricks. But persuade yourself that you have mislaid the purpose of these actions. You have no idea what lighting a fire means. You don't even know why you have always found it strangely moving, attractive, and reassuring.

So, you will accomplish each action exactly as you have always done, but this time with your full attention. Here you are on your knees, or squatting down, in front of the hearth. You make sure the ash is well-bedded or pushed back, that there is enough space and ventilation. Don't start with logs that are too big. They should be slightly raised as well, resting on endirons, bricks, or, failing that, on bits of wood, so the air can get beneath them. For the same reason, take care not to stack your wood against the back of the fireplace. Now add a sufficient amount of kindling—twigs, sticks, the wood from fruit and vegetable crates. Tear pages of old newsprint into strips (use newspapers rather than magazines, which burn less well) or twist them into ropelike shapes. Don't block the airspace; leave some room underneath the logs.

Set light to the paper nestling right in the center. What happens next is the most interesting part. There is an immediate response, cracklings, sudden tongues of flame, a few loud bangs. Next, the wood starts hissing as the vapor is expelled. And then there's a moment of doubt and anxiety once the paper has been consumed: the flames disappear, the embers are almost nonexistent— only a dense smoke gives some room for hope. The smoke keeps coming, dense and fairly thick. The burnt paper crinkles red around the edges, and then goes out completely. You suspect that the fire will never take, that you've made a mistake (wood too damp, paper badly crumpled, too tightly or too loosely). Everything is as usual, in fact, and yet a doubt remains. Absurdly and un-justifiably you start to fear failure.

You are uncertain whether the red flush on your cheeks is due to the heat or to your impatience. Unsuc-cessfully, you blow on a few sparse embers. The smoke gets thicker, the hissing more intense, but the fire still doesn't take. You are almost resigned to having to add more paper and start the whole operation again. You hes-itate. Suddenly, through the smoke, which they disperse instantly, come small, livid, intense flames. As if the fire has suddenly taken. And now you watch as it starts to bite and lick at the bark on the undersides of the logs, how it honeycombs them with red. All is well.

Ask yourself what it was that caused you such anxiety just now, and reassures you at present. Perhaps some atavistic memory from those gloomy times when losing fire was the only hell. Each sudden leaping flame confirmed that magic victory over night, hunger, cold, and death.

46 Be aware of yourself speaking

Duration: *a few minutes*

Props: *none*

Effect: *disconcerting*

Usually we follow the thread of our ideas, thinking about what we want to get across, not about the form of words, or their pronunciation. If the latter starts happening to you, it's difficult to keep going, and the experience is an unpleasant one. Like speaking into the phone and hearing the persistent echo of your own voice.

In fact, we never need to be aware that we are in the process of speaking. Either we remain silent, or else the effort of finishing what we have to say takes up nearly all our mental concentration. It is not advisable to say to oneself: "I am in the act of expressing myself, I am speaking the sentence that I am speaking." Thinking this way, we run the risk quite simply of paralyzing our ability to speak at all. Such awareness must not be allowed to impinge on our speech, our lecture, our political oration, or on any occasion, indeed, when the public has every right to expect our sentences not to stop dead without explanation.

Ordinarily, each of us manages to avoid this kind of derailment. The trick is to keep the knowledge that we are ourselves speaking in its place–which should be minor, marginal, and unfocused. Cling to meaning, to intention, and to everything the sentence refers to beyond itself. Keep going and don't look back or over-prolong a pause–or you risk blocking the whole machine. Speech can only go forward.

What curious conclusion is to be drawn from this situation? We can only speak on condition that we ignore the fact of our speaking. That we draw a kind of veil over its self-evidence. We can, of course, talk about language itself, but that's not the point. We cannot speak and think about speaking at the same time.

47   *Weep at the cinema*

Duration: *about 90 minutes*

Props: *a film*

Effect: *soothing*

The film must be the right kind–unintellectual, easy to follow, and predictable. A love story, nothing else will do. Sit a bit too close, so as not to miss a moment, melt into the screen, forget everything. Believe that everything you see is true and noble. Utterly sad and utterly beautiful. Turn into a softie, a romantic schoolgirl. Completely, otherwise it isn't cinema. Abolish all critical distance and solemnity. Rid yourself systematically of all resistance and disbelief. Determinedly, resolutely, become a devoted public.

And when the lovers part, when the heroine dies, when murder, evil, or stupidity triumph, when dreams are shattered and hearts broken, when the violins surge in a minor key and the drums resound, just cry. With large, warm tears. Without thinking and without being ashamed. Hotly, intensely, interminably. Feeling despair and comfort at the same time, carried away by the story, incapable of the slightest resistance, ravaged by grief, happy to be able to weep, and to think of nothing else.

Because we live in cynical times, cold, critical, sneering times, it's good to experience feeling in a free and accepting way. Without ulterior motive—just for pleasure. Flaunting this orgy of apparently innocent tears conceals a particular pleasure, that of letting go, of barriers temporarily lowered.

48

Meet up with friends after several years

Duration: *2 to 3 hours*

Props: *a friend from the past*

Effect: *chronological*

This is an experiment that can be carried out at different ages, and it takes on a different coloring depending on where one is in life. Children who haven't seen each other for two or three years scarcely recognize each other, even if they are old enough to remember and have played a lot together. "It's Anthony, isn't it? Do you remember him? Yes, you do . . . It's Marion, look, she's got the same eyes!" Embarrassed smiles, avoidance of eye contact. They'll get to know each other, but remember nothing, or almost nothing, and the little they do remember is a blur.

In adolescence, when one hasn't seen a friend for a long while, the experience is both amusing and disturbing. The mixture of amusement and uneasiness comes from the fact that features remembered clearly are now accompanied by breasts, and hair, and everything else that is growing in those years. You see your smaller selves in grown-up bodies. And that's bizarre.

Adults can go for longer. Ten, twenty, thirty years without seeing one another. The anticipation in the restaurant or the café, wondering if you'll recognize each other and in what guise, with what wrinkles, all the wear of time. It's a rare mixture of apprehension and tenderness, and you scarcely know whether the apprehension is for yourself or for the other. The tenderness likewise.

And then the strange incredulity with which—having recognized the other instantly (how? the eyes? the smile? the angle of the head?)—you scrutinize those ravages of time. The other person has aged, clearly. You have too, as you're aware, though you can't see it. And then, in a strangely moving way, time's terrible passage hits you with a sudden dread, obliquely, as you suspect that you, also . . .

49 Browse at the bookseller's

Duration: *2 or 3 hours*

Props: *several second-hand bookstores*

Effect: *distracting*

It happens by chance. Unexpectedly, you have a little time to spare. Between two meetings, or because there's a train strike. Or you happen to be passing by. Whatever, it shouldn't be premeditated. There you are in the book-shop. Never mind the town, the district, the country, or the season. Chain store, independent, used, the only thing that counts is that you enter completely into the whirligig of titles.

You go from shelf to shelf, stack to stack, wall to wall. You're not looking for anything in particular. And you feel solicited by titles, authors, characters. As if each book were calling you, trying to attract your attention. Each cover is like a closed window with the shutters up, and behind each one you sense whole lives and souls. Within each volume destinies await you. Insignificant destinies, passing dooms. No matter. If you enter you'll be carried far away, and for a long time.

But the competition is intense. Among these thou-sands of books, which will you choose to follow? You can make out the murmurings of these titles as they tout for your attention: "Will you read me, sweetie?," "Take me with you, you won't regret it . . . ," "If you open me you won't be able to stop!," "You're the one I've been waiting for! Take me, take me!" You pass quickly from one whisper to the next. Texts, texts, offered up to you, you hear their low voices and breathe their warm breath.

And the truth comes home directly: literature is prostitution. At least in one sense. Each printed story is a hooker trying to be noticed, trying to captivate the passerby and live a little longer in the arc of your attention. All the arts are like this: works, whispering their obscenities in a low voice, as your gaze slides from one to the next.

You may end up equating bookshops with brothels, exhibitions with group sex, and culture with an orgy. Even when your free time comes to an end, they call after you. And you feel profound compassion for the artists.

50 *Become music*

Duration:	*20 to 120 minutes*
Props:	*a piece of music*
Effect:	*realist*

Turn your amp up full volume, or very nearly. Put on some music you like and close your eyes. Relinquish any kind of vigilance. Make no effort to hear or think anything whatsoever. Relax every muscle in your body. Your body drowses heavily, languidly on the sofa (or the floor, or the bed) where you are stretched out. Everything else dissolves.

Nothing but the music. Wait, and let go. Any effort on your part is useless and even negative. Music and nothing but, absolute, imperious, and alone. It does not invade you, it is you who are dissolving, into notes, rhythms, and timbres. And you know full well how inexact and approximate a formula that is. Words aren't made for this. You must wait for that wordless moment when you cannot even say you are floating in music because, for such an expression to have meaning, it would still imply a "you" separate from the sound itself. It is precisely that "you" that is canceled. Nothing remains but sound waves, that pure pulsation you have become.

You can now do the experiment–fugitive, oblique, extreme–of looking at that abandoned, inert body from without, and from a distance, without worrying that it's your own. All those stories of shamans and sorcerers, who leave their bodies for a time, and look down on them, are in fact stories about music. How come you never knew this before now?

But it was still only a stage. Magic, special powers–these are moments to be journeyed through. Becoming music is useless for travel: if one persists, the idea of moving from one place to another does not even arise. There is no here, and no there. Soon there are no landmarks left at all. Only the music survives. It is the very texture of presence, the direct access to existence. And now you understand that once incomprehensible statement: "If the world were to disappear, music would remain."

51 *Pull out a hair*

Duration: *3 seconds*

Props: *a hair*

Effect: *minuscule*

The pain is negligible. A pinprick, a tiny, brusque removal from the scalp. You've got hold of one of your hairs between finger and thumb, and pulled it out, in a sharp, vertical movement. Possibly you hesitated, thinking it will hurt more than you expect. Or possibly it took several attempts, before you found the single, abrupt movement required.

And now you've got it. The hair is between your fingers. On your head, at the exact spot where it was, you feel a stinging, which radiates outward in motionless, but growing, circles. A pain both specific and vague—and unusual in that it starts out in a single, circumscribed spot, and then starts to fade, until it is indistinct. A memory of pain extended over the skin, rather than something you still really feel.

Well, that's a pretty stupid experiment, you say. Completely useless, utterly without interest. You are absolutely right. And that is exactly the point: to make us aware of an infinity of questions that have no point and no answer. There you are, less one hair. How many did you have a moment ago? How many have you now? Did the exact number of hairs on your head never concern you? Why should it not concern you? With one hair less, have you become bald? How many pulled hairs makes a man bald? Who knows?

These are questions that have no answer because they touch upon the frontiers of our being, the limits of our identity. These frontiers are not straight lines: we know when a person isn't bald, and when he is, but we have no instrument to gauge precisely, to the nearest hair, the line dividing the bald from the not-bald. We discern our own presence in a similarly approximate way. What is clear is that we are vague assemblages, auras, mists–incapable of knowing, to the nearest hair, quite who we are. We shall continue to know nothing of this, and consider the situation no more painful than a pin-prick.

52 Walk in an imaginary forest

Duration: *2 to 3 hours*

Props: *a forest*

Effect: *astounding*

Preferably, it should be a real wood. Winter is best, or a season in which walking briskly and for a fairly long time does not make you too hot. Your breathing should be absolutely regular. So there you are, walking at a brisk pace, for a good spell, without attending to anything except the synchrony of your breathing and walking.

The first stage consists in generating, by means of repetition, a completely regular, almost somnambulistic rhythm. It is easy to verify if you've reached the right tempo: stop walking suddenly, and continue to breathe at the same rhythm. If the trees keep moving forward, then you're on the right track. But if the landscape halts when you halt, keep walking; you haven't yet reached the beginning of this business.

When you've got into the correct rhythm, keep going. You are about to enter a different country. You do not have to discourse with fairies or elves, gnomes or trolls. All you need is a bit of goodwill, and a certain perseverance. A pinch of enthusiasm.

Imagine now that the forest is your soul. You are walking like this within yourself. The tangle of tall trees, the white sentinel of the birches, the moss and the damp mulch, none of all this is outside you. Some obscure spell—which is none of your business—has turned everything inside out. You are strolling through the inside of your own thought. You have a niggling feeling that maybe we are never outside ourselves. Don't ask why that might be so. Just take note that these semi-tones and half-lights are inside you. They belong to you, as intimate parts of yourself, along with the shadow in the undergrowth, the serenity of the clearings, the long-weathered tree trunks, the transparent lightness of the airy spaces.

You start to see that the mind has nothing outside itself, or if it has, then we can know nothing of it.

Up to you to draw out the weighty consequences of this sylvan game. It may be enough to retain just one ground rule: the imaginary is never, and should not be, something added on to the real, and which opposes, contradicts, or dissolves it. Reality itself must always be rendered imaginary.

53 Demonstrate on your own

Duration: *30 to 40 minutes*

Props: *an open space*

Effect: *depoliticizing*

You are strolling calmly down the street. But it's only an appearance! Even though no one could tell from your walk or your movement, you are in fact on a demonstration. It's your secret. No one could so much as guess at it. No banner, no chanting. There isn't the slightest clue as to anything unusual about your progress. Nothing, absolutely nothing appears out of the ordinary.

So you walk along in silence. But you are chanting slogans in your head. Slogans very hostile to the government and its policies. Pithy, catchy, sardonic one-liners. And then you move on to insults, defamatory and libelous remarks. You challenge authority, you stand up to the police, you gather opinion, you proclaim your determination, you protest.

But nobody knows anything about it, not the woman you pass, or the child who overtakes you. Not even the cop observing you indifferently from the street corner. This is not a demonstration. You are alone and silent. Just an experiment to find out. But to find out what?

That it's possible for everyone to do likewise without anyone knowing. The peaceful street, where people come and go, minding their own business, could be the silent theater of secret protests and invisible uprisings.

In fact, it is. Think, for a second, of all the grand passions that mingle unknowingly on a banal and neutral street. Since walking along the pavement, in the last few minutes, you have passed a terrorist, a woman with cancer, a jobless depressive, a junkie in need of a fix, a pregnant schoolgirl, an illegal immigrant, a broken heart. And you knew nothing about it. You couldn't know anything about it. Demonstrably.

54 Stay in the hammock

Duration: *precarious*

Props: *a hammock*

Effect: *risky*

It must not be wide. Discount those flat-bottomed hammocks that are really just suspended mattresses. They have nothing in common with the true hammock, the recalcitrant, unstable, authentic hammock. Unless you were born in one, and learned since childhood how to sleep and turn about in one, first attempts can be risky. Sudden, spontaneous movements are to be avoided. The best way is just to let your own, languid weight do all the work.

Once inside, you need to bear in mind that no stability is achieved once and for all. A fall is always a possibility even if you think you've got the knack. The best way to keep your balance is to remain aware that it can be suddenly upset. At any moment. Without your knowing why. An upset is always possible and the fact has to be accepted, made light of, and kept in mind but at a distance. There is no way of preventing a fall except by remembering that it is a permanent possibility, and keeping half an eye on it.

Mastering the hammock is, in fact, a way of discovering how extremely heartening the practice of pessimism can be. That the worst may always come to pass, but not necessarily, actually gives a rather sunny cast to reality. You find yourself rid of the illusion of certainty, cousin to the dread of disappointment. Maintaining a supple equilibrium is to be recommended.

To recapitulate. A fall is always present as a possibility, but it is no more than that. Accepting the risk becomes a means of protection. Maintaining a certain irony vis-à-vis the worst. Clearly, we must master life as we master the hammock.

55 Invent headlines

Duration: *about 15 minutes*

Props: *pencil and paper*

Effect: *calming*

You are cut off from everything. It happens. Not even a radio or a telephone. No TV, no papers. Completely disconnected. And yet you need your fix of news. Specialists claim that our craving for news takes different forms, some more intense than others. Some of us need an injection of news several times a day. Others are happy with a headline or two, morning and evening. Headlines can be taken in pill form, diluted within a news program, or straight off the screen in "breaking news." You can get your fix by fax, e-mail, or the Net.

But this time you are completely deprived. No gadget at hand and not a house on the horizon. All the same, you are going to have to make do. So you are going to invent your own headlines. They can't be transmitted to you? Well, we'll have some just the same! Don't worry, it's not as difficult as all that. For home news, there'll be a choice between a politician's resignation, a new raft of measures (dedicated, depending on your mood, to taxation, education, transportation, or the environment), a scandal, a compromise, a polemic, an official visit. In the foreign section there'll be a war, a coup, a meeting of specialists (again, depending on your mood, it'll concern monetary matters, the electronics industry, or fishing rights), a terrorist attack, an earthquake, a fire, a flood.

Do not forget the latest scientific advances: one step closer to human cloning, revelations concerning a traffic in human organs, a new way of storing data. Throw in a bit of culture: the latest films, a new exhibition, a profile of a writer. Carry on, if you feel so inclined, with a few celebrity titbits: an actress divorces, two princesses wed, a singer arrested for speeding.

Now for the final flourish: a few *faits divers*, a rape in the country, a murder in the suburbs, a bus overturns on the highway. That's it. You've just about got all you need, barring a few accessories. But there's nothing to stop you fabricating more, a weather spot, the stock market, the lottery. If you feel there's something missing in your line-up, invent a death: of an eminent politician, a Nobel Prize–winner for literature, a celebrated film director, complete with life, works (to be praised), and place in history.

Killing time is not the point of this experiment, which is rather to prove to yourself how the flood of news never ceases to repeat itself, and how it is always the same. It shows neither progress nor novelty. The extreme ease with which it is possible to fabricate pseudo-news only goes to confirm that there is nothing less new than the news. All it shows, interminably, is the endless misfortune of man. To add a little savor to this admittedly rather dismal message, try to come up with something really extraordinary. If you do this experiment from time to time, you might conclude that these tons of information are really of no great importance, almost without reality, in fact. Is that news to you?

56 Listen to short-wave radio

Duration: *15 to 60 minutes*

Props: *a short-wave radio*

Effect: *cosmopolitical*

This is old-world magic. In the age of the Internet and the Web, the radio has something prehistoric, obsolete, almost pathetic about it. It recalls those old-fashioned materials like Bakelite. Radio is the product of a different era, and it shows. Nevertheless, its relative antiquity need not stop us experimenting with it.

Get hold of a short-wave radio. The experiment is best done at night; the reception is better and the hallucinatory effect is always greater (though there's nothing to stop you carrying it out in daytime). Switch on the radio, and start turning the tuning dial, slowly and without stopping. Don't try to find out what you're listening to. If they exist, ignore the names on the dial and the cursors. Whether you're listening to Helsinki or Madrid, La Paz or Toronto, is of no interest here. Remain in the dark and keep tuning. The magic still works.

You pass, second by second, from one universe to another. The timbre, speed, and intonation of the voices change. Some languages you recognize immediately, others are harder to identify. Is that Hungarian? How do you tell Bulgarian and Romanian apart, if you know nothing of them? And what about the Scandinavian languages, the languages of Asia? You feel both very close and very remote from these strangers all speaking at the same time. You can hear them distinctly enough, some of them as if they were in the next room. Yet you don't know where they are, what they are saying, or even the name of the language they are speaking.

You advance surrounded by speaking shadows and absent presences. You know that they are alive, in all probability, but without being able to say where they are, or anything about their lives. You can imagine them as your fancy takes you, in a studio speaking into a microphone, in a neutral space here, a wretched one there. And you can imagine their listeners, Serbian peasants, traders from Cairo, executives from Copenhagen–inhabitants of dissimilar places, dressed differently, and sharing neither the same tastes nor the same terrors.

And so you stumble against one of the enigmas of technology. In solitude and in silence, in the most splendid isolation imaginable, perpetually around you, woven into the air, impossible to detect without a machine, are these hundreds of voices murmuring, in dozens of unknown or unrecognizable languages, of which you know nothing, except that they spread an obscure and changing human crust, unendingly, over everything.

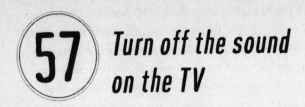

57 Turn off the sound on the TV

Duration: *about 5 minutes*

Props: *a working television*

Effect: *instructive*

We do not, as a rule, really watch television. Even those who spend a lot of time in front of it, consuming program after program, do not really observe it. What we are almost always doing, in fact, is listening. Sound and image are given whole, they make sense together. Whether the program is stupid or sublime, we take it en bloc, we listen, and we see. We do not scrutinize the images in their strangeness.

Start by cutting out the sound, and watch. First of all, you'll be invaded by a sense of the ridiculous. There is something risible about them, those people debating, gesticulating, and getting heated for no reason, the presenters who wriggle and smile and pout, pout and smile and wriggle, with those fixed smiles before the credits, before the next one comes on. Those voiceless singers, mute journalists, dumbshow actors who articulate without uttering a sound and scream in silence; the ads without music or enthusiasm.

But there's worse to come. Underneath the ridiculous, at varying depths and densities, lies terror. Something mechanical, fixed, and inhuman inhabits these faces, vainly moving their lips and filling their cheeks. Something quite different from death, the coldness of corpses, their white immobility. Instead, an agitation emptied of life, a vain effort to escape the void, a movement that keeps canceling itself.

But you must get beyond that too, beyond the horror and beyond the comedy. Stop laughing and stop frightening yourself. Move forward gingerly between the ridiculous on the one hand and the mechanical on the other. Look at them for what they are, these soundless images: at bottom they are insipid, savorless.

Neutral and empty. Even TV can lead us to wisdom.

58 Rediscover a childhood scene that seemed larger

Duration: *instantaneous*

Props: *a childhood scene*

Effect: *dislocating*

It happens to us all. In memory, which is often precise and detailed, it was huge. A vast field, an endless esplanade, an enormous courtyard, an expanse of rough ground to run about in, a place for hide and seek, for ambush, a pampas, a steppe, a virgin forest. Crossing the middle of it, without cover, took courage. To go from one end to the other demanded patience, and a fair amount of strength.

We didn't grow up there. And we never got the chance to go back and measure ourselves against it as we grew. And now, as an adult, we are suddenly there again. It's a tiny space. Reduced, shrunken, stunted, shriveled away. And yet it's just the same: the proportion of the whole, and in the details–that corner window, that little patch of faded yellow on the wall. It's a scale model, a maquette, a monument in miniature.

Experiment with this feeling of surprise and strange unease. You take yourself as a reference point, but you have no immediate perception of your own change in size. So you imagine, spontaneously, that it is the things themselves that have changed, shrunk, and crowded together. Which is why, all of a sudden, you feel like a giant.

Even though you know full well that this is not the case, your unease remains. Your memory and your present perception refuse to be synchronized. Each is vivid. Both are incompatible. You feel in yourself a false relation, and somehow excluded, between your memory, which is correct, and your current perception, which is correct too. What encumbers you is the notion of your own continuity.

59 Get used to eating something you don't like

Duration: *a few years*

Props: *a detestable food*

Effect: *civilizing*

You really do not like it. There's nothing to be done—having tried to get to like it, honestly and after several attempts at different intervals, you still—well, can't stand it! You were not ill after swallowing it, nor did it tickle or irritate you. So you are not allergic to it. It's purely a matter of taste. Nonetheless, you're going to eat it.

First of all, try the odd mouthful from time to time, in a stoical manner, and in a spirit of independence. And then a little more often, more regularly, more straightforwardly. After a few years of dogged persistence, you'll be eating it almost without noticing. You still won't like it. But the fact will have become banal, your displeasure blunted, and a state of indifference almost attained. You might even end up with a predilection for the food, not because of its taste (still hateful, though miracles can happen), nor because you've grown used to it from repetition, but out of a kind of tenderness toward yourself for having overcome your original aversion.

You may well ask why anyone should choose to impose such a constraint upon themselves, especially one that seems gratuitous, probably stupid, and unpleasantly pointless. The reply will silence all dissent: in the name of civilization! For in what does civilization actually consist, whatever the place or the period? In not following blindly everything that attracts, in not reacting mechanically against anything that repels. Civilization arrived to complicate the issues and to check those impulses. The answer to your question must be that the aim of this long experiment is to help you play your part, in a spirit of education and disinterest, in humanity's great adventure.

Only cynics will see in this argument a perfect apology for barbarism.

60 Fast for a while

Duration: *12 to 36 hours*

Props: *none*

Effect: *empty*

It is no accident that every spiritual tradition everywhere in the world and in all ages has used fasting. Drugs apart, no method has more power to alter our relation to the world. To frighten or to soothe, to compel or to cause indifference, to disturb or to calm—fasting is capable of provoking all these reactions, depending on time, intensity, and circumstance. None of this should come as a surprise, incidentally—our most ancient, constant, and organic relation to reality is via food. To abstain from eating voluntarily, therefore, goes against the very bedrock of our being.

Everyone reacts differently, depending on their past and their internal architecture. One may be filled with dread at the thought of entering an arid, mineral landscape, a desert of stones that no sweet taste will ever soothe or relieve. Someone else will welcome it, by contrast, pleased to be released from the constraints of having to eat, freed from the ghastly obligation to ingest solids at set times.

If you have no experience of fasting, try going for a single day to begin with. Doctor's orders apart (and these are rare) it is not dangerous. Take some sugared water at regular intervals. Don't expect anything amazing. Just explore your mood changes, and more generally what appears to you to be "reality" after just a few hours.

A little less glucose, a bit less fat, some missing proteins, and you no longer see the world in the same way. Don't you think you should draw out all the possible inferences of this discovery? Well, try.

61 *Rant for ten minutes*

Duration: *as above*

Props: *none*

Effect: *bracing*

There's always a pretext. You are angry and in a bad mood because you've been hurt. Someone has upset you. There's been some unpleasantness. So you rant. Right or wrong (absolutely right, in your opinion). You protest against an inconvenience or an injustice. You show your displeasure by screaming and grinding your teeth.

The game consists in doing the same thing, but doing it cold. Gratuitously and for no reason. You are going to act out the gestures of anger without feeling anger. On your own, in a room, you start to rant for no reason.

Get beyond the initial embarrassment. Start growling and try out the appropriate noises in your throat. Contract your diaphragm. Bark and curse, scream out the ready-made phrases with violence: "It's unbelievable!," "I cannot believe this is happening!," "How dare they do such a thing!," "It's disgusting!"–"Bastards! shits! assholes!"

Do not think of anything at all. Beware of provoking any genuine flash of anger in yourself. Just utter the words. Remain calm. Keep going. Imagine you are being filmed and that you must look convincing. Keep ranting. Stamp your foot and beat your fist, or both, as the mood takes you. Scream that it's disgusting, odious, hateful, inadmissible. You'll get them. They'll pay for this. You'll crush them, reduce them to pulp, force them to beg for mercy, they're going to regret this. Make a few more loud noises in the back of your throat. Make ample use of your breath and your glottis.

Then stop. Breathe. Drink a glass of water. Open the window. And remember that a temper tantrum might in fact be nothing more than this.

62 *Drive through a forest*

Duration: *10 to 20 minutes*

Props: *car, forest, road*

Effect: *Jurassic*

The trees flashing by produce, almost immediately, an effect of unreality, like a film. If you are lucky enough to be a passenger, turn your head to the roadside and keep your gaze fixed on it, to watch the trees go by, mechanically and at high speed. That is just the first stage in this hypnosis. The second is attained by watching the undergrowth, which is dark or less dark depending on the forest. This time keep your gaze fixed on the gaps between branches, on the green light, which sometimes looks almost black. How would you live in this place? What would it be like to be here always? As it might have been in the Middle Ages, here or anywhere else, before they started clearing the forests on a grand scale.

The car affords you special protection. You speed on through, protected by glass and metal. But the subtle terrors of the forest still get to you. Even if you accelerate, you realize, suddenly, that you'll never get out of the woods.

63 Give without thinking about it

Duration: *instantaneous*

Props: *at hand*

Effect: *gracious*

The boredom of repeated actions—the dreary tasks of a predictable routine—has completely invaded you. You move forward mechanically through a gray mist. Almost indifferent. To yourself, as well as to others and to things. Then some wretchedness rises up before you. Real wretchedness. A tramp sitting there, a ragged child, a cripple, who is ill, skin flushed red by the cold, by nights sleeping rough, by drink. Street corner, pavement, subway exit, red light. There, all of a sudden.

Give them something immediately. Without knowing, thinking, calculating. Without deliberation. No theories, no justifications. Give what you have at hand, a note, a sandwich, a book, a pen, a smile. Pay no attention to the value of the objects or the appropriateness of the gift. Experience the act of giving wholly. Not exactly anything to anybody, but give a part of yourself, who are relatively protected, relatively comfortable, and not in any real need, to the other. Just because a hand is held out to you.

In the cold light of retrospect you may find your action to be arbitrary, and in a way unjust, because of its unreasoning, contingent nature. As soon as you start to think about it, you'll sure enough arm yourself with every good or bad reason, and conclude that your alms-giving was useless, beggary immoral, and charity suspect. Which is why you should block out with violence any thoughts like these that come to afflict you at the moment of giving. The gift springs from pity, from the sudden impulse to give help, and to show solidarity—an impulse that is countered and dissolved by analysis.

In these circumstances, you should not hesitate to shut your reasonings up with the use of force. The gratitude shown in such moments remains in the memory. What you have long since forgotten are the little meals digested properly, and the money spent prudently, logically. But what you do remember are those you gave away. And the gestures, faces, and words that went with them. It's not just the reverse of forgetfulness. It's the opposite of remorse.

64 Look for a blue food

Duration: *indefinite*

Props: *indeterminate*

Effect: *imprecise*

Seen from space, we inhabit the blue planet. It's true that there is, on earth, a phenomenal amount of blue. The sky in daytime, when unclouded; the oceans. We are ceaselessly immersed in blue, we see and breathe it. But we cannot eat it. Blue is inedible. It escapes our devouring.

This is a very straightforward and yet a very considerable mystery. Foods exist in all colors. Almost all of them can whet our appetite. But nothing blue can be eaten, and the presence of a pale blue food, or even one in ultramarine, runs a good chance of looking repulsive. The sight of royal blue icing sugar gives one the impression not only of extreme artificiality, but can also provoke a kind of indefinable malaise.

A few very rare exceptions exist. They are not especially convincing. Stilton or Roquefort frequently shades into green or black. Blue Curaçao belongs among those sad, fake-tropical, lagoonlike cocktails. The liqueur known as Lorraine Thistle has fallen into a discreet desuetude. The Vosges "Blue Line," an old French specialty, is almost entirely forgotten. In any case, it was only ever consumed by the eyes.

So, you can always go on looking. There is nothing blue to eat. Or at any rate, not commonly and with appetite. Not like green, red, yellow, orange, even black or white, which are all copiously ingested. What are we to make of this? That we can't digest the sky, the ocean, the planet itself? Need one recall that blue is also associated with royalty and with death? Mysteries, nothing but mysteries!

65 *Become a saint or sinner*

Duration:	*15 to 20 minutes*
Props:	*none*
Effect:	*relativizing*

Are you good? Are you bad? A host of consequences
hangs on the answer. Such is the common belief, at any
rate. Because we invest the question with meaning. A
brief experiment can easily convince you that the question, so grave in appearance, has no foundation.

Consider how you spent yesterday. Retrace the main
events, how one led to the next, and, as far as possible,
reconstruct the thoughts that went with them hour after
hour. From this reconstruction, consider your attitude.
Try and judge it. Not objectively, as if you had managed to
achieve an external and neutral view of yourself. On the
contrary, proceed insidiously. In a partial, exaggerated,
and tendentious way.

Note first the extreme magnanimity of your smallest
actions. Be a benevolent judge of your innermost thoughts.
Look how devoted you have been, how attentive, altruistic, sympathetic, disinterested, modest, efficient, humane,
supportive, charitable, respectful, the whole day through.

Perhaps, at first glance, it does not bear too much scrutiny? The aim of the experiment is that you should come to consider your words and actions in this light. Never mind what you really did accomplish. You must end by picking out–from the day before–the manifest signs of your saintliness. When you have the feeling that this judgment is pretty much secured, replay the film.

And then do exactly the reverse. Force yourself to discern, in your acts and thoughts during that day, the obvious signs of your perversity, your ability to harm, your taste for destruction, your fundamental wickedness. Once again, neglect the apparently banal unfolding of your behavior and your feelings. Find, in everything that was said and done, confirmation that your character is ignoble, spineless, hateful, diabolical, cruel, egotistical, and manipulative. See yourself as an arch-sinner. With no more justification than seeing yourself as a saint, but, if possible, with no less plausibility either.

Apply the same test to the people around you.

And then, if you have carried this out completely enough, try believing in moral judgments and the searchings of conscience.

66 *Recover lost memories*

Duration: *30 minutes or longer*

Props: *a memory*

Effect: *unpredictable*

You know so many things! The fact seems so simple and obvious that you are not even aware of it. However learned or otherwise you may be, you know thousands and thousands of words in your mother tongue, rules of grammar, arithmetic, and geometry. You can also recall countless stories, real or invented, lived by you or those close to you, transmitted by historians or witnesses, taken from tales, novels, films.

The experiment consists in showing that you hold in your mind more memories than you yourself know. Naturally, you know that already. But do the experiment all the same: spend half an hour in an armchair. Close your eyes, and go in search of a lost memory, in a deliberate and willful way.

You might think that a hunt of this type is bound to fail. Too direct, too crude. Not at all. It rarely fails. Nearly always you end up with a memory to explore—a fact, a date, an action, a scene, a face—that you thought had gone forever.

Don't set off completely at random. Start with a few major categories: work, holidays, family, or historical events. Decide upon a starting point, and a guide—a face, a year, a place, an emotion. Follow, explore, dig down. Don't force it, scarcely use any willpower. Let it come. Just when you are about to give up, a fragmentary image surfaces, a sound, a smell, a scene. Sometimes you receive it whole and all at once. At other times you have to open it up, unfolding its edges one by one.

From time to time pass up the chance of a walk outside. Rather than going out, go in, wander about, browse within yourself, and uncover, as one does a mushroom or a truffle, a memory invisible on the surface.

67 *Watch someone sleeping*

Duration: *a few minutes*

Props: *someone sleeping*

Effect: *endearing*

You are familiar with each centimeter of her skin, the timbre of her voice, the movements of her eyes, and nearly all her reactions. You love her laugh, the carriage of her head, everything, even (for instance) that tiny imperfection, possibly known only to you. In short, you are already acquainted.

However, if you watch her sleeping, you will very probably have the impression of not knowing her completely. The face is no longer present to itself; there seems to be an inner absence. Eyes closed, body languid, posture unusual, that ineradicable innocence. And her breathing sounds far away. So why do you feel this curious mixture of total confidence, slight anxiety, and vague embarrassment, as if you were present at some scene you should not be witnessing?

Most likely it is the juxtaposition of presence and absence that creates this unease. Perhaps you cannot really be sure that this Sleeping Beauty is indeed the same as the woman you love. You will never know. It may seem droll. And yet. All you can do is carry her along with your tenderness, which you can extend as far as possible, into the living heart of this silence, of which she knows nothing.

68 *Work on a holiday*

Duration: *8 hours*

Props: *a national holiday, a job*

Effect: *socializing*

That's just how it is. From force of habit, and because everyone does the same. Most probably it is not an inclination, still less a pleasure. But the habit is formed. You have gotten used to it, whether your work is educational, professional, or domestic. Sometimes you have to keep working, while everyone else is having a day off. While they are sleeping, dreaming, walking, doing odd jobs, at the cinema–you have to work.

The experiment consists in observing closely what happens and what you feel. After a few spasms of ill humor, a feeling of being duped somehow, of vague persecution, of generalized frustration at the order of things, you emerge into a kind of suspended space. Nothing is as usual. You do the same things as ever. You carry out the same tasks. But the air seems different. The background noise is missing, the general hum and effervescence created by others working alongside. Obviously, all the objective signs are there–the telephone is silent, the offices empty, the streets relatively deserted. But that is neither the most interesting, nor the most enigmatic aspect of all this.

There is, in fact, no verifiable clue to the fact that other people are not working that day. Especially so if you work from home. Nothing indicates that this is a day off, and yet you feel it in a tangible, in almost a palpable, physical way.

Does the difference exist solely in your imagination? Or is there in fact a form of collective perception, a social sensation, a subtle sense of everybody's noise? Decidedly, life is full of mysteries.

69 *Consider humanity to be an error*

Duration: *about one hour*

Props: *none*

Effect: *bracing*

How often we have been told that we are exceptional! Center of the world, children of God, universal consciousness, salt of the earth, intelligence, language-beings, spirit of science, vector of progress. Our existence has been so celebrated by myths, religions, philosophies, smug ideologies, that it is hard to comprehend our failures, our vileness, our interminable wars and our endless filth. Naturally, there has been any amount of special pleading, to explain our fall, our malediction, and our two-facedness.

There is a way of experimenting with a more radical form of disillusionment, which is doubtless more beneficial. Rid yourself of anything that resembles a meaning to our existence. Consider humanity as a result of pure chance, a mistake, a biological accident. It developed without order, on some lost pebble in some small benighted corner. One day it will disappear forever, unremembered and unmourned. For the tens of thousands of years of its survival, our species stagnated. Then it multiplied unreasonably, and plundered its own habitat. And before disappearing, it will have charged to its account a weight of suffering both unimaginable and futile, massacres and famines, enslavements and tyrannies.

Take a clear-eyed look at this absurd and violent species. Confront its lack of justification and its ephemeral, irrational existence. Train yourself to endure this vision of humanity as fundamentally meaningless and futureless. This should contribute to your serenity. For against this background of unmeaning and horror, every sublime thing shines out the more as a matchless gift. Perfect music, unforgettable paintings, the glory of cathedrals, grief-stricken poems, lovers' laughter . . . Such are the endlessly surprising fruits of this aberration that is us.

70

Inhabit the planet of small gestures

Duration: *variable*

Props: *a past*

Effect: *migratory*

Believing in the existence of a single world is both crude and deplorable. A fly's universe has nothing in common with a whale's, or with your own. It is doubtful whether these multiple universes intersect in any way. What you call your own world is made up of a considerable number of distinct planets, not necessarily connected.

To feel this, try to inhabit the planet of small gestures. Banish from your memory everything having to do with music, sound, color, shape, taste. Try to replace them with memories of movement, tactile sensation, the traffic of small gestures. Some will have marked you, for sure; they remain there, buried under the rest.

Here are just some of my own: a hand on my forehead, placed there by a woman on a day of great distress; the unforgettable way another woman put her arm in mine, in a public garden, one autumn day; my father stroking the nape of my neck; the way my mother used to make, when she said goodbye, a curious sign with her hand.

You know your own. Find them. Notice how they are linked, and how they weave a different world. The network of small movements has an independent existence. Explore the planet they make up. Its traffic is not the same as when you are navigating your other memories. And from time to time it's worth your while spending time there. You move by a process of contiguity, from one gesture to another, like a circuit, a series, markers on a path. Besides, it's impossible to get lost.

71 Disconnect the phone

Duration: *variable*

Props: *a telephone*

Effect: *ambiguous*

You like your friends to call you, to know that your family can reach you, that your clients ask for you, and that your colleagues keep in touch. We are all pretty much alike in this. But sometimes we may feel disturbed, and then irritated, by a ringing phone–the way it can intrude at any moment, interrupting an activity, wrecking a conversation or meditation.

Leave the phone off the hook. Switch off the mobile, pull the plugs from the walls, make sure nothing can ring. For as long as need be. Don't busy yourself too quickly, don't hurry to make use of this moment of guaranteed quiet. Before you rush off to do some work, or to take a siesta, before you fling yourself into a domestic chore or a sensual pleasure, stop a moment, and taste the feeling procured by this retrenchment.

Sometimes it brings real satisfaction: out of reach at last, left alone, and free to get on without interruption. Sometimes it creates anxiety: what if there were an emergency? A really serious piece of news? An accident? At other times it creates guilt: there are people trying to reach me, and they can't even leave a message, just because I prefer my own comfort to their demands—can I justify this?

It may seem a kind of revolt, a tiny rebellion against the age of permanent connection. It has become so normal to be online, so indispensable to be connected, that cutting the wire can appear like closing a door and leaving, a first step beyond social control, a move into elemental liberty. Concomitantly—and you feel this too, instantly—it's a step back into the wild, an asocial act, which brings with it a piercing solitude. And you wonder what to do. Why not telephone a specialized helpline?

72 *Smile at a stranger*

Duration: *minimal*

Props: *none*

Effect: *complicity*

In the street, in shops, at work, at the market, in the village or the town, abroad or in your own country, you do not know the people you encounter. Often enough you've never seen them before, and you'll never see them again, especially in a city or at a tourist attraction. You may well have no desire to communicate anything to these strangers. You have the most fundamental right to remain closed, mute, indifferent, or frosty.

Try smiling. Discreet, restrained, clear but reserved, purely benevolent. Try. When your eyes meet a stranger's, when you are alongside each other for a few seconds. It is not always easy. If it's too emphatic, your smile can seem idiotic or equivocal. Too subtle, and it may pass unnoticed. One has to find—and it differs according to circumstances and individuals—a smile that says: "Let's try to show a little tolerance toward one another, and since I've no reason either to resent or like you, and vice versa, I wish you a good day." Or else: "Let's be indulgent. Stop." Or anything else that takes your fancy.

There's no reason to invest this experiment with any particular virtue. It may strike you however that its adoption generally could soften our manners. If it doesn't increase our hypocrisy. Or both. Though perhaps an improvement in our manners is not necessarily a good thing. Unless a little more hypocrisy is in fact desirable. And suchlike questions that raise a smile.

73 Enter the space of a painting

Duration: *undefinable*

Props: *a painting*

Effect: *displacement*

On the whole, the spatial organization of the world holds no surprises for you. You have the map. Whether it's a matter of localizing an object or of judging a distance, you are perfectly at ease. If we leave to one side the exceptions—remote bodies and intergalactic paradoxes—in daily life and in your immediate surroundings, space is regularly organized. Free of snares or booby-traps.

Unless you fall into certain paintings. It's impossible to say in advance which paintings will trigger this effect in you. So you have to try the experiment in a museum of your choice, without knowing what will happen. Let your gaze slide over the smooth canvases. They may be interesting, stirring, cleverly composed, masterly, sublime—but they still exist in the same space as you. And then, if you're lucky, suddenly something else happens.

You feel attracted, drawn, called by a kind of fault within normal space. This fault belongs to an "impossible" space, to another dimension, to a hollow in the texture of the world. There are in fact several types of space into which you can fall. Some look like crypts, others like staircases, still others like inverted basements or perspectives in spiral. Others are striped with infinite lines, or punctuated by black holes. Or like melon rinds, or pumpkins, or Cheshire cats.

Above all, do not hesitate. Go with that initial attraction. Let yourself be seized, wafted, carried away. There's nothing to be afraid of: we never return from these spaces. We remain within them, without ever leaving our own. So we exist, permanently, in several spaces. Which is why the arts intensify existence.

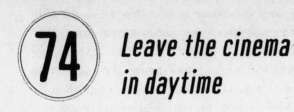

74 Leave the cinema in daytime

Duration: *about 90 minutes*

Props: *cinema auditorium, daylight*

Effect: *disjointed*

You have followed the heroine, the struggles, the reversals of situation for a long time now. You've lived in the dark under different skies. The cinema has emptied you of your current preoccupations and filled you with its images. It has washed you of time and its uniformity. You reach the corridor or the staircase that leads outside. Under the lights, you recover some portion of the normal world. But as yet it's only a passage, a moment of transition. And then you open the door.

Outside, it's sunny. You had forgotten that. Totally. You wonder how it's possible—not your forgetting, but this daylight. *Ext., day*. It wasn't in the script. It ought to be dark outside. As it usually is, with the rare passers-by, the speeding taxis, and the shop windows all dark. But no. It's very bright, and this hurts your eyes a bit. There are lots of people on the pavement—what have they been doing all this time? Working? Walking? How have they gone on existing?

Very well, they must have managed as usual. But still. Their dogged persistence seems rather strange, and even something of a provocation. When you're caught up in the same tide, when you work or take the bus with them, you don't even notice it. You know they are just going about the business of existence. But now, while you are still with the heroine and her struggles, you cannot imagine how they've coped.

Their comings and goings have simply gone on as usual. Their time is not out of joint, their movements are uninterrupted. Not yours. For you, duration has become distended. It has made a large pocket that has contained the plot of the film, its landscapes, your emotions, and perhaps whole lifetimes. Fairly quickly, the whole question fades and then disappears. But only by neglect or distraction. It is never really settled.

75 *Plunge into cold water*

Duration:	*1 hour and 20 seconds*
Props:	*sun, sea, sand*
Effect:	*transitional*

If your health permits, start by remaining a good while in the sun. One hour at least, on a very hot day. You must feel your skin cooking, and even your insides growing warm. Choose a time and a place where the water is particularly cold. The sharper the contrast between hot sand and cold water, the more successful the experiment.

Run abruptly into the water. Without stopping. Jump or dive straight into the cold water and swim under the surface for as long as you can. Those few seconds are worth exploring. You don't feel the shock of the cold immediately. The first tenths of a second radiate with pinpricks, or sparks striking over the lines of your body. Like a hot and cold shower, going into free-fall, an effervescent instant of transition. And then, a second or two later, the cold hits, that breathtaking grip, and with it the need to keep swimming, underwater, going down deep, before at last resurfacing to take a gasp of air and emerge into the white glare of the sun.

Those for whom prudence excludes this game can explore a similar sensation by different means: jump into a bath full of lukewarm water, or, if you are particularly fragile or timid, plunge a foot into a basin of icy water. The interesting thing, every time, lies in the stupefying contrast produced, and the sense of being out of synch with oneself. It feels as though such disjointed sensations, which are so remote and distinct from each other, cannot be embraced all at once within the supposed continuity of the self. Their disparateness would seem to exclude them from being contained in the same cycle. Arriving too fast in succession, they leave the self like a tortoise on its back.

76 *Seek out immutable landscapes*

Duration:	*interminable*
Props:	*the Earth*
Effect:	*perennial*

It is not exactly nostalgia. A certain tenderness, perhaps, a slightly melancholic form of curiosity. It drives one to seek out landscapes that are the same today as they have been for tens of thousands of years. Are there places that haven't changed at all? That bear no trace of human activity?

What forest has remained the same? What region, landscape, hill can show a wholly unmodified face? What mountain, even? You start looking. You try various approaches. Approximations and hesitations. There's always a nagging doubt: hasn't agriculture changed everything? Erosion? You may imagine that a particular panorama, taken in the round, has remained identical to what a Stone Age man might have seen. But you're never completely certain. Which leads to disappointment.

There is a solution, close at hand. Put out to sea, until you can no longer see the coast. Nothing has changed here. An identical stretch of water is still there. From time immemorial. What you see was seen by pterodactyls. And it still accounts for almost two-thirds of the globe. In other words, the greater part of the Earth has remained unchanged. Alongside the catastrophes, the earthquakes, the changes wrought by man, the greater part of the planet has retained almost exactly the same appearance, wet and blue, as far as the eye can see.

Draw whatever conclusion you like: a matter for amazement, an object of controversy, a reassuring fact, or a bitter disappointment. The foam endures.

77 Listen to a recording of your voice

Duration: *a few minutes*

Props: *a recording of your voice*

Effect: *dislocating*

It is always a shock. "Is that really me?" Your own voice seems too high or too low, too slow or too fast, unsteady, inappropriate, out of phase, unexpected. At first both the timbre and the delivery seem unrecognizable. And yet the recording reproduces the voice of other people faithfully. Not your own, however.

You know perfectly well it was you who uttered those words and sentences. And you recognize your own speech immediately, but at a slant, in profile, from a strange angle. You and not-you. You fall into a gap, into a void that has suddenly opened up. You are familiar with yourself from "within." Now you perceive yourself from "without." Professionals are used to it. People in radio and recording know their voices from within and without. They work with and within this substance. Used to hearing themselves, they feel nothing of the surprise and unease that other people habitually experience when they listen for the first time to their voices as others hear them.

In the old days, no human ever heard his voice as others heard it. Just as they never saw their images as others saw them. Machines have rendered this decentering possible. Not that this is an escape from the self. Rather, our instruments confirm that our knowledge of ourselves is made up of ignorance. Here, technology is an aid to philosophy. It prompts us to ask which image we should retain: that of ourself as we receive it from within, or that which seems objective and can be recorded? It's the same for our face, for our thoughts, for everything that makes up our general behavior. There has never been an absolute answer to this one. And that always comes as a shock.

78 *Tell a stranger she is beautiful*

Duration: *less than 1 minute*

Props: *none*

Effect: *fireworks*

You have never seen her before. Pure chance and timing account for the fact of glimpsing her today, in the restaurant. Or on the train, in the café, crossing the street. She is radiant, alive, perfect. Her very presence is uplifting. In a few minutes, or a few seconds, she will disappear. You will never see her again. That is of no importance. You are filled with gratitude for her brief stay. You want to thank her for existing, to tell her she is beautiful and that her beauty rejoices the heart.

This is simply not allowed. You run the risk of being misunderstood. If she is alone, she'll think you are simply trying to pick her up (even though your gratitude is disinterested). If accompanied, you will be cast in the unpleasant role of the lewd provocateur, someone who deserves to be slapped.

All the same, dare yourself to do it. Out of style and sincerity. You have more to gain than to lose. Gain what, exactly? The pleasure of saying it. You cannot thank a landscape, a flower, or a bird for the joy they procure you in contemplating them. They know nothing of that moment of recognition that beauty can cause. In the human case, it is different.

As to what happens next, you will find out for yourself. But if the response is in most cases a shrug of the shoulders, this at least testifies to a deplorable decay in the social fabric.

79 *Believe in the existence of a smell*

Duration: *between epsilon and infinity*

Props: *none*

Effect: *canine*

You think you have smelled something. A rather sweet smell. Like a vague perfume that you cannot quite place. A breeze of flowers, a touch of spring, a memory of incense maybe, or the smell of young hair. Don't try to identify it. Keep the trace of it. Enter into the fragrance, however slight, attenuated, or evanescent. Amplify it, nurture it, draw it out, and keep it going, and let the world contained within it unfold.

Is the smell really there? Or have you just imagined its existence? It's of no consequence. You can always extend your trust to a passing smell, welcome and develop it, without asking whether it is true or false, good or bad. Welcome the different smells of a country, a house, a person, a situation, the smell of fear, love, death, childhood, school, work, the kitchen, the market, smells of all kinds. You will come to see how this neglected olfactory universe, sometimes scarcely considered worthy of attention, actually lies midway between the real and the imaginary. Here, the two are difficult to distinguish; they change places or even interpenetrate each other. The kingdom of smells is a halfway house between dreams and wakefulness, the palpable and the illusory.

Raising unusual questions, odors are the subject of unresolved scholarly debate. For example: does the smell of water change with temperature? Can one smell the odor of an odor? Does the essence of sandalwood (for instance) smell the same as the existence of sandalwood (and so on)? Is there, or is there not, an odor to the world? And furthermore: are people who suffer from anosmia (having no sense of smell) atheists? Ought they to have the same social status as blind people or deaf-mutes? Keep going.

80 Wake up without knowing where

Duration: *5 seconds*

Props: *a room somewhere else*

Effect: *cosmopolitan*

You need to be fairly exhausted. Or else to lead a stressful existence, traveling about a great deal. You are asleep, and suddenly a noise, a light, or just the alarm pulls you out of it. You know you are not at home, but, for a brief lapse of time, you don't know where you are. Five seconds is the usual interval. Then you remember, you see, you know: the town, the house, the why and wherefore. The experiment consists in exploring this moment, suspended there between your first awakening and when you find your bearings.

Despite its brevity, it is a moment of considerable interest. You feel, effectively, as though you've slipped your moorings. Weightless. Not necessarily worried, or preoccupied, but surrounded by whiteness, by pure light. As they do in movies, you can say: "Where am I?" In actual fact you entertain no doubts as to the reality of the world, and the continuity of your own existence within it. What you do not know, for a tiny interval of time, is what the place is called, where it is, and what you are doing there.

But you're certain that you are somewhere, and will find out where very soon.

Which confers on this brief episode the delicious lightness of a mystery without menace. Your query is a real one, but it will soon be answered. Your ignorance is not a pretense: you really don't know where you have woken up. At the same time you repose in your sovereign knowledge of the world: you are somewhere, there's no doubt about it, and in a trice you will remember where, oh yes! that's it, of course. Try not to lose hold of this rare moment of perfect suspension between doubt and confidence, certitude and ignorance, anxiety and satisfaction.

81 Descend an interminable staircase

Duration: *a few minutes*

Props: *a staircase of 8 to 20 flights*

Effect: *introspective*

For the first two or three flights, you are just getting into step. Establish a regular rhythm as soon as possible, synchronizing your feet, the movement of your legs, and your breathing in such a way that they become automatic, and you can keep going without thinking about it. When you feel a slight dizziness, you've reached the right mechanical state of mind. Keep going.

Imagine that you'll keep going down like this forever. This spiraling movement will continue indefinitely. Bottomless: no hell, no material disintegration, no death. Just this regular, interminable descent. As far as the eye can see. And there's no stopping. If you wish, you can embroider your downward journey in different ways. Imagine you pass through different zones of color, for instance. Invent frozen flights, torrid zones, light and dark flights, long overcrowded passages, deserted intervals, parts of the staircase better maintained than others, local populations, folk music, regional dishes, rustic paintings. What you can never alter, however, is the fact that your descent has no end.

Rely on your gallows humor to compose two funeral orations in your head, with an elegiac warmth: one for the inventor of the ground floor, the other for the salesman of staircases.

82 *Swallow your emotion*

Duration: *variable*

Props: *none*

Effect: *calming*

We tend to forget that the ideal among the best of men, for many a century, was to rid themselves of their emotions. Being able to shrug off that chaos, dump that load, stifle those harmful fires–these were noble tasks. Romanticism, it must be admitted, has not helped us in this forgetting. It turned emotions into a huge adventure, and treated them as signs of a great destiny. Under its influence, they have become glorious, and sometimes grandiose. Enviable, at any rate. The classical age, the inheritors of Antiquity, did everything they could to marginalize them. Emotions were to be banished.

Examine the ideal of the Sage in Antiquity. If the Sage is happy and to be admired, it is because he has freed himself from the tyranny of the emotions. He lives beyond their reach. He ignores them, having become impermeable to their existence. The Sage is never moved.

You will probably never be a Sage. But try all the same to swallow an emotion. When it arises, refuse to go along with it. Regard it as you would a boil, an inflammation, a temporary swelling, and will it to go down. Try to look at it from without, from where it seems both laughable and unpleasant. Don't enter into the emotion. If you find yourself inside it, find a way out. Press it down. But don't get too involved with that task either. Let it go.

Naturally, this is sometimes very difficult to do. If you are tormented by some anxiety, submerged in some dread, transported by some joy, swallowing emotions promptly and completely is sometimes scarcely possible. But not impossible. The important thing is to set yourself an ideal. Either a life without sadness or panic, without heat or enthusiasm, a storm-free existence, or else a life made up of contrasts, implosions and explosions, terrors and ecstasies, laughter and tears. Try experimenting with a foretaste of one, then the other. Or else, if you can, invent an alternative. Humanity will be forever in your debt.

83 Fix the ephemeral

Duration: *stopped*

Props: *a recording system*

Effect: *contemplative*

In the past, the fleeting detail was lost forever. A movement of the hand, a furtive look, the shadow of a smile, an inflection of the voice, a certain light–everything sank without trace, along with a thousand other such tiny realities–into an ocean from which nothing could be recovered by anyone.

We have invented detail-machines that can capture such instants. Fix the tiniest spot. Preserve sounds and profiles. They made their appearance not so long ago, but we have got used to them so quickly that now we take their existence and their power for granted. Or nearly.

The experiment consists in rediscovering the power of these techniques. In fact we do so every day without thinking about it: by listening to the radio or to discs, watching television or videos, photographing, recording voices, faces, images, etc. This time let us think about it– about how these machines, in their very special way, can snatch ephemeral moments from the march of time.

A piano key lightly caressed by Scott Joplin in 1902, a single flutter of Louise Brooks's eyelids in 1934, a jack-boot hitting a paving stone on the Champs-Elysées in 1940, the chaotic arrival of a train in Calcutta yesterday . . . We are surrounded by tens of thousands of captured living instants, of movements seized in photos or on film, of every type of music recorded on disc. Try to remember this when you next look at a photograph or listen to a CD. And the next time you yourself record a snatch of life, remind yourself that you are seizing a tiny sliver of existence from time. Think of a paradoxical hourglass that sieves the dust of hours into a kind of eternity. Instant after instant, what would have disappeared forever is now available and can be replayed any number of times. The fugitive and the ephemeral are flowing permanently toward the eternal.

Ask yourself how all this will end. The answer doesn't matter.

84 *Decorate a room*

Duration: *intermittent*

Props: *some rooms*

Effect: *adjustable*

Wallpaper and carpet, tiling and paintwork, electrics, light fittings, doors and windows, cushions and curtains, furniture, plants . . . You have to adjust their arrangement, their color, and their style. The interesting thing is that one doesn't know quite how to go about it. Learn to listen to what the room tells you. Each place calls for such and such a shape or arrangement. It's impossible to have a total or rational knowledge of the matter. As if the spirit of place spoke its own language in each room, which you have to learn on your own. Hence the need to steep yourself in the character of the place: volume, light, surfaces, textures, grain. After which it's trial and error.

No decent home decoration springs fully formed from a first impression. There have to be approximations, and you have to proceed step by step, prepared to make mistakes. Learn to renounce and rediscover, work below words and signs. Not in any way abstractly or theoretically. You try out one color, and all the surrounding colors change. Install a piece of furniture, and the space changes, and sometimes the color and light as well. Everything reacts to everything else, always. Which is why, properly speaking, you ought never to get it wrong, even if you don't exactly know the right procedure.

The experiment, therefore, is continually subject to different rules. You must learn both to act and to let be. You set to work, of course, but the thing might succeed just as well if you imposed nothing. On the other hand, the consequences of this relative passivity will depend on who you are. What any particular place seems to dictate or demand will never be the same from one individual to another: if you let yourself be guided by the place, it is still you who are being led, not somebody else. Thus it's not just the room you are decorating bit by bit, but yourself too.

The experiment shows us to what degree we are party to our own surroundings. Quite unlike an author, or a designer, who bring an external act of will to the business of appearances. You yourself become a constituent element of the room, just as it constitutes one of the elements of who you are. If someone says "nice place you've got here," you can either take it as a thoughtless banality, or reflect on how truth and its effects are processes.

85 *Laugh at an idea*

Duration: *unpredictable*

Props: *an idea or two*

Effect: *encouraging*

Can ideas make people laugh? The complicit smile, the amused reaction–these we know. What about the real laugh, the full-throated guffaw? Laughter is not discreet. It should really be neither polite nor in good taste. A human trait, it is also animal, uncontrolled, improper.

If you really want to experience laughter at an idea, you should frequent philosophers. There you'll find countless notions–unusual, bizarre, cooked up, contrived, twisted, counterfeit, deformed, grotesque, eccentric, aberrant, lunatic, hilarious, farcical, appalling, frightening. Moreover you'll see these notions thought through in squintwise fashion, from the side, at a slant, in profile, in suspension, in the air, from behind, with eyes closed, without hands. Among the solemn, dogged, minor thinkers you won't find many such acrobatic displays and magic tricks. For these, you have to turn to the great, the true, and the authentic. It takes real geniuses to make us laugh, clearly. Because they can invent rarefied concepts and notions that leave us open-mouthed on our first acquaintance with them—proof-shaking machines and thoughts that, at first sight, seem unthinkable.

At first you may hesitate to guffaw at the good, the true, and the beautiful. So start with the oddities. Search in hidden corners. For example: Plato on dogs, Aristotle on the erection, Spinoza on tickling, Pascal on sneezing, Kant on ants in the Congo. This is a good way to get acclimatized, but it's still too easy. It leaves intact the belief that there are the oddities and curiosities on one side, and the true, eternal, and legitimate questions on the other. Quite the contrary—you should be able to laugh at Plato's idea of the Good, at Aristotle on the *primum mobile,* at Spinoza on Nature, at Pascal on the God of Abraham, and at Kant on moral law. Among others.

To reach this stage, you need not only time, a bit of reading, and a little patience. Above all, you must learn to abandon that tiresome habit of dividing the important from the laughable, solemn-visaged respectability from grotesque hilarity-unleashing farce. The heart of the matter is a laughing matter. So try to get rid of the idea that to laugh at the greatest ideas is somehow to despise them.

The truest way to respect ideas comes through laughter. Explain and discuss.

86

Vanish at a pavement café

Duration: *30 to 40 minutes*

Props: *café, pavement*

Effect: *diaphanous*

It should be full of people. Almost crowded. At last you find a place, a little table at the back. Order, wait until you're served, and then ask the waiter if you can pay right away. Whether he pretends to notice you or not is of no importance. In three seconds you are going to vanish.

You feel no particular discomfort. You still perceive things pretty much as normal. And yet, since the moment the waiter gave you your change, you have become transparent. No one can see you. All around you people are talking, but none of them speaks to you. They look straight through you. Nobody comes to occupy your chair, but that's just chance. There you are, absent, dissolved, untouchable. Suddenly obliterated. Undeniably real to you, your existence has become invisible to others.

You can escape this critical situation by resorting to the well-tried expedient of the invisible man: spill something that stains over yourself. But it isn't foolproof. Especially as the same thing can happen to you in the subway, during a play, or anywhere, in fact, where no such solution is available. In the end, the thing is always in doubt. You can never be absolutely sure that, seen from the outside, your existence represents a continuity. There are of course numerous and compatible clues that may allow you to affirm that others do see you, speak to you and, on the whole, do not mistake you for another. But nothing proves that it is always thus. Quite the contrary, there are many circumstances in which you can remark your own absence in their eyes and behavior. Usually, it's enough to say something, ask a question, or mark your presence in any way, to dispel any lingering doubts. It doesn't work every time. There are a certain number of situations in which nothing can convince you that you have not disappeared.

Bearing these moments in mind, the question is to know whether you are worried or reassured by this state of affairs.

87

Row on a lake in your room

Duration: *about 1 hour*

Props: *rowing machine or boat*

Effect: *symmetrical*

Occasionally you engage in the act of rowing, either on a rowing machine at home, or, indeed, in a boat on a lake. But did you know that rowing is a philosophical activity? Several features combine to affirm the truth of this. Probably of negligible significance is the fact that the familiar French expression "I rowed like hell before getting there" designates a painful effort with no direct result. More pertinent, on the other hand, is the notion it contains of moving about ingeniously on an element that is not naturally ours. Notable also is the fact that it explores surfaces, exterior spaces, and has no access to what is underneath, or out of sight. But this is still not the determining feature.

What links philosophy most closely to a rowing trip is the fact that, in both cases, the movement required is continuous and its regularity is the product of a–necessarily fragmented–succession of repeated efforts. Each pull on the oar is independent of the others, each one, singly, drives the boat forward, which advances in a jolting movement. But when the oarsman executes these successive thrusts smoothly, the boat moves ahead continuously and steadily, without any jolting. This is certainly a legitimate image for philosophy: several individual, discontinuous thrusts that in the end act together to create a uniform, uninterrupted trajectory. The key to the process? A judicious use of inertial force, glide, and breath.

There's one essential rule to rowing: a respect for symmetry. At the slightest imbalance—caused by pulling harder on the right than on the left, or vice versa—you change direction, the boat rocks, your back tenses up, everything goes wrong. You must, therefore, do everything to keep this symmetry and not break it. You should even extend it to your mental images. So the next time you're on the rowing machine in your room, imagine you're on a lake, look at the color of the water, respect the rhythm and the symmetry. And the next time you're in a boat on the lake, look at the furniture in your room all around you, and keep the rhythm and the symmetry.

When you have sufficiently linked in your mind these two ways of rowing, to the point when doing one evokes the other, ask yourself what relation all this has to philosophy. Continue for as long as it takes to see the answer clearly.

88 *Prowl at night*

Duration: *2 to 3 hours*

Props: *a city, night*

Effect: *nomadic*

A lot of people think that night can be defined as the absence of solar light. Such a conception is not only simplistic, it is false. Night gives us a new planet, quite other than the one we know by day. Codes and movements are not the same. Thoughts are different. And it's by no means certain that individuals remain identical with themselves.

This strange mutation is undoubtedly at its most visible in cities. The night has its own specific population, which is intermittent, rarefied, and dispersed. The night is a space, first of all, of different dimension from that belonging to daylight. People prowl within it. Which means that they move around with no specific aim, unless it be in search of some kind of prey, which is also difficult to define. The streets are almost empty. It is possible to proceed unhindered, and to see the shape of the city. Whether one is on foot, or in a car, the city at night lies wide open, inviting us to explore it from end to end.

Try taking your car, or your courage in both hands, and just head off in any direction, for a large part of the night, without knowing where you're going. You will always come across something new, even if you become a regular nighthawk. A particular neon light, a district, a fight, some horror or other, a spectacle, a run-of-the-mill orgy, an empty quarter, a spontaneous party. Above all you will find that there are multiple nights, divided up by sectors in space and by phases in time. You may come to wish that the night was endless. You will look on the dawn as a defeat. You will live in the anticipation of twilight. The end of the day will be full of strange promise.

You will question the truth of this definition: "We call the philosophers of the Enlightenment those who are drawn, out of the darkness, by everything that shines." You will inquire what relation this might have to the fact that in ancient Greek the plural of the word *psyche*, which means the soul, refers to butterflies.

89 *Become attached to an object*

Duration: *several years at least*

Props: *any ordinary object*

Effect: *consolidating*

It is not its beauty or its value that matters. It may be a perfectly ordinary object. Not really decorative, or useful. Almost of no interest. You kept it out of laziness, almost by chance, as if it had been stuffed in the corner of a cupboard. Or because a member of your family had given it to you—one of your children or your parents. Or it's a love token, from some affair now over. Or a souvenir brought home from a journey, something to help you remember an episode or a place. To begin with, you did not pay much attention—it was not an object you loved. You may not even be too sure now where it came from, to what or to whom it was linked.

Since then this second-rate object has traveled down the years. Without your ever really choosing, it has ended up among the things that were never thrown out, sold off, or given away. It has followed you from place to place, through your variously tortuous wanderings. Then, one day, you remembered where it came from. You recalled who, or what, it was linked to, and since then you have never forgotten. And now a kind of fondness, born of long habituation, has overlaid the original object. You feel a kind of gratitude to this object for the mere fact of persisting.

By now, even if you are not in any way a fetishist or otherwise superstitious, it has become precious to you. You would be angry were it broken, and sad if you were to lose it. You have an affectionate, lasting, and finally a reassuring relationship with this object. One day, perhaps, when you are very frail, sick, old, dying, or some day when you feel everything collapsing around you, and you are falling helplessly into a bottomless pit, you will take this old object in your hand and cling to it as to the only thing that still holds together and prevents you from going under completely. Who knows?

90 Sing the praises of Santa Claus

Duration: *about 10 minutes*

Props: *an audience*

Effect: *reviving*

Ceasing to believe in Santa Claus passes for a proof of maturity. You have passed the age of gullibility, those tender years when you could be made to believe anything you were told. You left that stage a long time ago. Now you are one of the grown-ups—a bit distrustful and disenchanted—but at least you can't be duped!

Are you so sure? And is it really so simple? Do you think this is nothing but a gain? You may be more autonomous, possibly more reasonable. But you are also less of a dreamer, with fewer expectations and a narrower horizon. Santa Claus, and everything that went with him, enabled you to preserve a few traces of an enchanted world. His fat, red joviality, his fabled gruffness, all this was steeped in fairyland. A tinsel paradise, a netherworld both accessible and reassuring. It is tempting, in our proud lucidity, to want to do without it. But it may not be entirely possible.

Santa Claus still visits, but in other forms, and less friendly ones, which claim to be more articulate, and apparently less naive. We go on dreaming. But this time in the name of science, of revolution, of success. And so we dream on, believing all the while that we've stopped dreaming. The man in red with his sled and his reindeer may after all be preferable.

So experiment with singing his praises in public, to friends and strangers alike. Lament the fact that Santa Claus has such bad press, that you hope he really exists, that you would like to see a great international inquiry set up, made up of objective experts, because a number of troubling facts—let's face it—still surround this affair.

You might remind people that Santa Claus is a benefactor of humanity, that for years and years he has brought dreams and toys to tens of millions of children. Stress that his reign dates from recent times, and that it is still fragile. In Dijon, in 1951, some Catholics burned an effigy of him, under the pretext that our gallant hero was a distraction to souls whose single preoccupation at that time of year ought to be the birth of Christ.

Argue your case with warmth, fervor, and conviction. What you really think doesn't matter. Be as convincing as possible. The point of the experiment is not to persuade anyone of anything. It is enough to observe that people's reactions, faced with your praise of Santa, invariably fall into two camps. Some will shrug their shoulders, consider you stupid or tiresomely provocative, and refuse to join in. Others will join in the fun, suggest that a committee for the defense be set up, and promise to have their chimneys swept. It won't all have been for nothing.

91 *Play with a child*

Duration: *30 to 40 minutes*

Props: *various*

Effect: *disorganizing*

This works best with a child who cannot talk, or who talks very little. Between one and two years old, for instance, and certainly under three. Choose a game he knows well, which he has mastered, in which he feels at home—and go along with it. Your role is simply to follow, and to join in. In his way, not yours. Accept the endless repetitions, the crazy rules, the delays, the inexplicable moments of merriment. The experiment consists first of all in entering this world of the child's game and leaving to one side, as far as possible, your normal adult universe.

The amount of effort and application required varies, as does the degree of suppleness or stiffness, depending on the individual: the goal is to become a part of this world of the child's game. Inevitably, your success will be partial. Nor should you become completely passive, but rather—however startling it may seem at moments—you should aim to become a true sharer of this world.

Afterward, explore the effects of this excursion on your return to the normal world. This is the heart of the experiment. If you have been sufficiently conscientious, if you have left behind, completely enough and for long enough, the thread of your own thoughts (incompatible with the world of the child's game), there is every chance that you won't be able to pick it up again immediately. What is interesting here—if you get to feel it—is indeed this moment of loss, of bewilderment, in which you grope around in search of your bearings.

As if entering this world of the child's game—be it for a brief and imperfect moment—left you so destructured that you have to work to recover yourself. This reconstruction can take a certain time. You may not instantly recover what you are meant to be doing, your hopes and your fears. You have been disorganized from within, and putting things back in their places requires a kind of effort you are unable to make on the spot.

You may care to meditate, in the wake of this experiment, on the narrow, and frankly fragile, territory you consider to be your "normal" mental state.

92 *Encounter pure chance*

Duration:	*2 seconds*
Props:	*a casino or equivalent*
Effect:	*adventurous*

You have just laid your bet. In a game of pure chance. No skill is required, and no intervention is possible. It works best if you are not indifferent to the outcome. What you stand to win must be substantial—enough, if you do win, to change your life in an appreciable way. So that you can say, with a degree of truthfulness, that the direction of your life is at stake. What you will be tomorrow depends, to a fairly large extent, on elements completely beyond your control: the trajectory of a roulette ball, the appearance of a card on a green baize table, the alignment of some images on the screen of a machine.

What you need to realize now is that this kind of situation is totally devoid of meaning. It is certain that you will win or lose. The probability of your losing is much greater. But the chances of your winning are not nil. Both possibilities are calculable. But neither has any meaning. This is the hardest thing to grasp.

From a financial point of view, you will remain as you are or you will become rich. But these divergent futures depend wholly on pure chance, without meaning or intention. The result has nothing to do with your merits or your faults. You are delivered up to the aleatory, to an arbitrariness that is both anonymous and without finality. In a way that has a sovereign disregard for any form of justice, you will become, in one second, a loser or a winner.

Inevitably you try to fill this void with all sorts of explanations, prayers, hopes, prognostications, and magic spells. Great strength of soul is required to accept that our life, in such a crude and naked way, be suspended like this, for an instant, over absurdity. If we were really great, we should doubtless be capable of it all the time.

93

Recite the telephone directory on your knees

Duration: *15 minutes precisely*

Props: *a telephone directory, preferably old*

Effect: *respectful*

People who like rituals never stop repeating them: perform the actions, and belief will come. Get down on your knees, go through the correct recitation, and faith will eventually follow. While it may seem offensive to those who have the true faith, the idea is certainly not without foundation. This much becomes clear from the following experiment.

For a few days, put aside fifteen minutes daily, always at the same time. During this quarter of an hour, read out loud an identical number of pages from the telephone directory. You will articulate clearly, line after line, names, first names, addresses, numbers. You will have taken care to seek out an old directory. This is not requisite, but it helps that this reading matter be as devoid as possible of any kind of utility, and assimilable to a venerable text, handed down through the ages.

Do not try to invest your recitation with any kind of meaning. Reject the idea that you are interceding with the telephone exchange on behalf of the departed subscribers, or that your prayer is contributing to the great universal interconnection. No. On your knees, every day, for fifteen minutes, you read out pages from the telephone directory. And that's it. This is what is known as a "practice." The rest is nothing but ornamentation and exegesis.

Besides aching knees, what will you get out of this experiment? A sense of the extreme force inherent in these absurd constraints, the strange fascination they exert, and the power with which, confusedly, one cannot help investing them. There is a strong likelihood that you won't be able to go on without furnishing some explanation. You will probably start by constructing a reason for your behavior. You will elaborate, if only for a joke, a myth that can incorporate your recitation, its meaning, and its goal.

If you find you can't stop, found a sect.

94 Think about what other people are doing

Duration: *10 to 15 minutes*

Props: *none*

Effect: *dissolving*

You are alone for a while. Rightly or wrongly, your solitude weighs on you. You feel cut off from the rest of the
world. False, of course. To find out just how mistaken you
are, begin by asking yourself what, at that very instant,
your loved ones, your family, and your best friends are
doing. Picture in your mind, as precisely as you can, their
current activity. Try and place them in space. In relation
to where you are now, are they in front of you? Behind
you? To the right? To the left? Higher? Lower? Picture
how far away they are from you. Look at them from various angles, as silhouettes and in the greatest detail. Now
widen your field of vision bit by bit. Admit to the scene
the people who are around each of them.

   Now think of what everyone else in that neck of the
woods is doing, in each village, suburb, city. Who is working, wailing, wandering? How many, at this moment, are
asleep? Consider the question from a planetary perspective. What percentage of human beings, at this precise
moment, are yawning? Cutting their nails? Writhing in

pain? Waking with a smile? Drinking soup? Screaming with pleasure? Dying of boredom?

How many, at this instant, are playing the piano? Listening to Bach? Fleeing the police? Leaving a library? On airplanes? Wiping their bottoms, washing their hands, brushing their teeth, blowing their noses? How many are crying and how many laughing? Speaking in front of an audience? Listening? Trying to commit suicide?

How many, at this instant, are asking the same questions?

95 *Practice make–believe everywhere*

Duration: *a few hours to a few years*

Props: *none*

Effect: *antidepressant*

The moments when you feel crushed are also those when you end up believing that life is serious, the world real, and words true. Countering this tiresome tendency is, luckily, not too complicated. It can be enough, at least when starting out, to transform each situation systematically, and turn it into a scene from a play. A metamorphosis of this kind affects not only your internal perception of events: it can change your voice, your gestures, your sentences, and even what happens.

This morning, for instance, you are not going to the baker's and then the post office merely to buy, respectively, bread and stamps. Start off by playing the triumphant customer, entering the bakery. Pay attention to how you thrust open the door (movement of the arm, full of energy, but not too brusque). Regulate your voice correctly, and sing out "Good morning," the greeting worthy of a customer who comes to buy a triumphant loaf. Ask, pay, and collect your change, triumphant still, say goodbye and thank you, watching your every movement, from

the confident step toward the door, to the little complicit smile addressed to the lady just coming in to buy—
inevitably—her sliced white bread and a bar of chocolate.

You now have three and a half minutes to put on a
different role, this time the anonymous purchaser of
stamps, who enters a strange post office, full of timidity
and apprehension; he doesn't know how things are done,
having lived too long abroad, or else he's just out of the
hospital, humbled in any case, with a guilty feeling, hold-
ing his loaf furtively in his hand, like an encumbrance or
an embarrassment, not knowing what to do with it, or
how to hide it . . .

And so on. Have a nice day.

96 *Kill people in your head*

Duration: *15 to 20 minutes*

Props: *none*

Effect: *relief*

From the point of view of public morals, the outlawing of murder is not a bad idea. But its drawbacks are not inconsiderable, when you think of the meanness and stupidity that continues to multiply with impunity. So if you find yourself raging against someone who is imbecilic, wicked, bad, or dreadful, don't hesitate—experiment with picturing his murder as clearly and distinctly as you can.

Choose the moment, the place, and the means (instrument, preparations). Review all your options with a clear head: if the murder is to be perpetrated by you yourself or by others on your orders, in your presence or not, with or without preliminary torture, with or without bloodshed. Picture the different scenes down to the nearest detail. Don't forget to decide what is to be done with the body. Spice it up, refine it, sharpen it, have no truck with sympathy or second thoughts. Go for the Grand-Guignol scenario, full of appalling gore and Sadean excess. Train yourself to consider these murders with satisfaction and to rejoice in them for a long time afterward.

Have no fear of awakening in yourself wicked thoughts capable of dragging you down the slippery slope of vice. Killing your neighbor in your head won't turn you into a criminal. On the contrary. The more you entertain this private pleasure, without a trace of shame or guilt, the more you will be able–having relieved your thirst for vengeance in pictures–to respect as you should the fundamental right of your neighbor, who is by no means dead, to life and health. The bastard . . .

97 Take the subway without going anywhere

Duration: *about 1 hour*

Props: *the subway*

Effect: *presence*

Public transport is excessively *functional*. You use it to go from one place to another. And that is all. Utility, and not pleasure, even if it transports you toward pleasure (you're happy to be going on holiday, to see the Alps again) or gives you pleasure (its comfort, anonymity, etc.). You are there to be transported, bundled, crated. You are not supposed to contemplate the scene, you are not meant to stand and stare.

Try it. Try being genuinely "displaced" by public transport, taking it but not using it. Strip the world of its instrumental function for a moment. For example, spend an hour on the subway, just for the sake of it. You can get on a train, take it to another station, make a change. The essential thing is not to be going anywhere, and that no particular predefined trajectory should give meaning to your being there. You have come to spend an hour, to have a look, and that is all.

There is a strong likelihood that this act of basic discrepancy will shed more light than you suspect on other people, on yourself, and, by extension, on the subway. For instance, ask yourself if it is really possible that you are, at that moment, the only person who is on the subway for no other reason than being on the subway. Might it even be conceivable that everyone you see, all those people getting on and off trains and generally carrying on like subway passengers, might all be doing what you are doing—that they are all here merely to watch the spectacle? If this hypothesis were proved true, then it goes without saying that the very function of public transport is just a ruse, an alibi for aesthetic souls, who are too timid, or hypocritical, to admit what they are up to.

Since this possibility cannot be verified or dismissed with any certainty, you are permitted to record your observations in a notebook that you will regularly update. In it, you can compare the many puzzles set by the subway networks of different capitals, European or otherwise.

98 *Remove your watch*

Duration: *unverifiable*

Props: *a watch*

Effect: *bewildering*

First, gauge the seriousness of your addiction. Do you consult your watch three times a day? Four times an hour? Much more often? You probably don't even know, and your guesses are wide of the mark. Begin with this objective test. If you consult your watch at least once every quarter of an hour, this experiment is for you.

It consists in living, in your "normal" life if possible, without wearing your watch. Start out for relatively short periods and when you are not racing against the clock. For instance, an afternoon at home on your day off. Then, little by little, try going watchless more daringly, when you go out, or to a meeting for work. Strictly speaking, you should refuse all chance succor, in the form of traveling clocks, kitchen clocks, public clocks, parking meters, or the digits showing the time on your computer. You should try on the contrary to settle fully into the slightly disconcerting queasiness provoked by not having the time on you.

You will feel the strange bareness of your wrist, and the mild dizziness of not knowing what has happened to . . . but happened to what? The sense of security provided by chronometers of all kinds? The alibis of accuracy? You feel, more or less intensely and more or less lastingly, a sense of unease. The world is not its normal self. It has slipped its frame, and floats, out of time.

If you persist, if you discipline and habituate yourself to going watchless, you should discover another form of perceiving time. Internal and organic, relaxed, exact but not neurotically so. You will end up by learning to tell the time by your internal clock, without needing even to think about it. This in turn may lead you to meditate on the particular–and completely relative–form of violence and constraint imposed on us by dials, clock-hands, and timetables.

99 *Put up with a chatterbox*

Duration: *a few minutes*

Props: *a conversation*

Effect: *restful*

They are prolix, unstoppable, and will not be put off. As soon as they have trapped you somewhere–in a doorway, a corridor, a reception, a dinner–they won't let you go. What they say is of no interest, but they are determined to tell you anyway. Chatterboxes are one of the scourges of humanity. But how to escape them? By learning to stop listening. To do this, you need a little practice. Most likely you won't manage it right away. A certain subtlety is needed, which can only be acquired. While the bore is speaking to you, as far as possible stop following the conversation. Under ideal conditions, you can contrive to hear almost nothing. With a little training, you'll end up not even knowing what the person with logorrhea standing in front of you can possibly be talking about. But the real challenge, of course, is to give no sign of this. The experiment therefore consists in absenting yourself, as completely as possible, but without your absence being noticed.

Never look away. On the contrary, look your bore straight in the eye, and put on the most attentive and interested expression you can muster (of subtle discreet amusement, or sad solemnity, depending on what you were able to grasp of the tenor of his initial remarks). Nod your head regularly. From time to time, punctuate by emitting that very short "a-ha" sound. But don't overdo it. Put yourself into automatic listening mode: you will follow, without the slightest effort of attention, the rhythm and music of this endless torrent of chatter. With practice, you learn to tell when a pause in the torrent is coming, which you can fill at random with remarks like "As much as that?" or "It's crazy" or "You amaze me." If need be, especially to close the conversation (or to change the subject before you go to sleep again) you can tune in to two or three sentences, and then ask a question.

Finally, try to perfect this art. The day will come when you can last a whole lunch hour with a bore who will be only too delighted by the interest you show him. Which goes to show how little it can take to be altruistic.

100 *Clean up after the party*

Duration: *1 to 2 hours*

Props: *a party at your home*

Effect: *variable*

The last guests have just gone. Everyone seemed happy. There were hilarious moments, warm hugs, old friends. Candles, jokes, games, music, and singing. Everyone brought something and, as usual, there was too much to eat. In short, it went well.

Now the house is full of dirty plates, half-empty glasses, and overflowing ashtrays. The kitchen is a scene of chaos, piles of plates tower alongside stacked-up cups, and the fridge looks as though it's been burgled. You have no butler, no housekeeper, and no maid. It's very late and you've drunk a fair amount. What should be done?

Two schools now confront each other at this moment, and they represent radically opposed conceptions of the world, and of our existential relation to time.

Proponents of instant action will tell you that you should get down to it *at once*, and plunge up to your elbows in the creams and the sauces, rinse all the utensils in hot suds, take out the bulging waste baskets, and put everything back in its place. Obvious drawbacks: having to keep awake and find the energy. Advantages: everything is clean the next morning.

Supporters of the take-it-easy school will send you off to bed without worrying much about the mess. Because they are not hygienists, they will argue that the chaotic mess after a party can give a special kind of pleasure, in all its joyous maceration and monumentality, and that cleaning up the next day will bring it all back.

These two schools are totally at odds. Their respective disciples have long since given up conciliatory talks. No one has ever, ever gotten them to agree.

101 Find the infinitesimal caress

Duration: *undefined*

Props: *the slightest*

Effect: *divine*

Caresses are moral. Because they are devoid of material-
ity. It is impossible to enclose them within a definition, or
in a confined space. A caress exists on the brink of its
own vanishing. Its mode of existence is a prolonged dis-
continuity, a sustained intermittence. It is endlessly re-
constituted on the edge of effacement, it quivers over the
void, it renders iridescent the borders of being. Applied
any more heavily, and it is no longer a caress. It gives
way to massage and stimulation, which are respectable
activities certainly, but of a different order. But without
touching, it cannot exist.

The experiment consists in finding the very lightest
possible caress that can still count as being one. The
lighter a caress is, in fact, the more powerful its effect
can be. And all the more exquisite for being so minimal.
On condition that it remain the lightest, the scarcest of
touches, at the edge, where being and nothing come so
close.

The infinitesimal caress is infinite in its effects. It is up to you to dedicate your existence to making at least a partial enumeration of these. In particular, you should compare the sensations procured from different parts of your face, your back, your tummy, your sex, making a note of the course they took. It is equally important to feel and ponder the differences between the tiny caress one gives to oneself and to another, and those the other gives to you.

Finally, you would be advised not to neglect the subtle links that exist between infinitesimal caress and ineffable ecstasy. They certainly constitute one of the frontiers of European history. On the one hand, the precept attributed to Him who, transfigured, was said to exist in a glorified body: *Noli me tangere* (do not touch me). On the other, the postwar Surrealist motto: *Prière de toucher* (please touch).